"十三五"职业教育国家规划教材

普通高等职业教育计算机系列规划教材

Android Studio 移动应用开发基础

吴绍根 罗 佳 主编

电子工业出版社

Publishing House of Electronics Industry

北京·BEIJING

内 容 简 介

本书是 Android Studio 移动应用开发系列教材中的基础篇，书中不仅讲解了 Android 的基本理论知识，还列举了大量示例来帮助读者理解和掌握知识点。主要内容包括 Android 概述、Android 界面开发、Android 数据存储、Android 多媒体开发及网络开发等。本书针对各个章节涉及的知识点，安排多个示例引导读者学习，由易到难，循序渐进。笔者通过一步一步的操作这些案例来介绍知识点的应用情况，同时，针对每个案例设计了对应的练习题，让读者在完成知识点学习之后能够有一个相对应的实践过程。

本书内容翔实，案例经典，实践性强，既可作为高职、高专移动互联应用技术专业课程的教材和教学参考书，也可供从事 Android 移动编程开发的用户学习和参考。无论是拥有丰富 Java 开发经验的程序员，还是只有 Java 基础的初学者，本书都将是十分有价值的学习资料。

未经许可，不得以任何方式复制或抄袭本书之部分或全部内容。
版权所有，侵权必究。

图书在版编目（CIP）数据

Android Studio 移动应用开发基础 / 吴绍根，罗佳主编. —北京：电子工业出版社，2019.8

普通高等职业教育计算机系列规划教材

ISBN 978-7-121-36993-3

Ⅰ. ①A… Ⅱ. ①吴… ②罗… Ⅲ. ①移动终端－应用程序－程序设计－高等职业教育－教材 Ⅳ. ①TN929.53

中国版本图书馆 CIP 数据核字（2019）第 131962 号

责任编辑：徐建军　　　　特约编辑：田学清
印　　刷：天津画中画印刷有限公司
装　　订：天津画中画印刷有限公司
出版发行：电子工业出版社
　　　　　北京市海淀区万寿路 173 信箱　　邮编：100036
开　　本：787×1092　1/16　印张：16　字数：461 千字
版　　次：2019 年 8 月第 1 版
印　　次：2022 年 1 月第 7 次印刷
定　　价：46.00 元

凡所购买电子工业出版社图书有缺损问题，请向购买书店调换。若书店售缺，请与本社发行部联系，联系及邮购电话：（010）88254888，88258888。

质量投诉请发邮件至 zlts@phei.com.cn，盗版侵权举报请发邮件至 dbqq@phei.com.cn。

本书咨询联系方式：（010）88254570，xujj@phei.com.cn。

前　言

本书是 Android Studio 移动应用开发系列教材中的基础篇，是一本介绍 Android 应用开发基础的实用教材，全面介绍了 Android 平台上应用开发各方面的基础知识。本书的主要风格是：语言通俗易懂、操作步骤详细、编程思路清晰。本书知识点的组织由浅入深，循序渐进，读者只要具备基本的 Java 基础，阅读本书就不会有太大问题。

本书中的示例都是针对知识点精心设计的，并且有相对应的练习题来供读者练习。在学习时，只要按照书中解决问题的步骤一步一步做下去，读者就会对所学知识点有一个清楚的认知，然后把书中对应的示例独立练习一遍，读者将会对知识点有进一步的巩固，从而做到"学中做，做中学"。

本书还具有如下几个特点。

1．基础知识全面

本书深入阐述了 Android 应用开发的核心基础组件，并详细介绍了 Android 全部图形界面组件的功能和用法、Android 各种资源的管理和用法、事件处理、Android 输入和输出处理、音频和视频等多媒体开发、网络通信等内容，其涉及的内容是所有 Android 开发人员必备的知识。

2．案例驱动，实用性强

本书没有枯燥的理论介绍，而是采用了"案例驱动"的方式来讲授知识点，每个知识点都可以找到相对应的参考示例，同时笔者还针对性地设计了对应的练习题供读者独立实践，使读者在学习之后能够通过实践再一次巩固知识点。

3．通俗易懂，讲解详细

只要读者具备一定的 Java 编程基础，阅读本书后就可以很轻松地上手进行 Android 应用开发。

笔者具有多年从事 Java 及 Android 等移动应用开发的教学经验。本书由广东轻工职业技术学院的吴绍根和罗佳组织编写，最后由罗佳负责统稿并审校。在编写过程中，企业工程师吴边等提供了大量真实的案例和许多宝贵的建议，在此，编者一并表示衷心的感谢！

为了方便教师教学，本书配有电子教学课件及相关资源，请有此需要的教师登录华信教育资源网（www.hxedu.com.cn）注册后免费进行下载，如有问题可在网站留言板留言或与电子工业出版社联系（E-mail:hxedu@phei.com.cn）。

由于编者水平有限，编写时间仓促，书中难免存在疏漏和不足，恳请同行专家和读者给予批评和指正。

编　者

目 录

第 1 章 Android 概述 ·············· 1
- 1.1 Android 是什么 ·············· 1
- 1.2 Android 应用程序的组成 ·············· 3
- 1.3 Android 的发展历史 ·············· 4
- 1.4 Android 应用开发环境概述 ·············· 4

第 2 章 建立 Android 应用开发环境 ·············· 5
- 2.1 下载和安装 Android Studio ·············· 5
- 2.2 开发第一个 Android 应用程序 ·············· 5
  - 2.2.1 创建 Hello World 程序工程 ·············· 5
  - 2.2.2 运行 Hello World 程序 ·············· 9
- 2.3 Android 应用程序的结构 ·············· 14
- 2.4 同步练习 ·············· 15

第 3 章 剖析 Android 应用程序 ·············· 16
- 3.1 AndroidManifest.xml ·············· 16
- 3.2 MainActivity.java——Activity 介绍 ·············· 19
- 3.3 Android 程序资源 ·············· 21
  - 3.3.1 字符串资源 ·············· 21
  - 3.3.2 布局资源 ·············· 23
  - 3.3.3 ID 资源 ·············· 24
  - 3.3.4 图片资源 ·············· 27
  - 3.3.5 Android 的其他资源 ·············· 28
  - 3.3.6 引用资源 ·············· 28
- 3.4 同步练习 ·············· 30

第 4 章 深入分析 Activity ·············· 31
- 4.1 Activity 的生命周期 ·············· 31
- 4.2 Activity 生命周期示例 ·············· 33
- 4.3 使用 Log 类输出程序调试信息 ·············· 36
- 4.4 Android 常见 Activity ·············· 38
- 4.5 同步练习 ·············· 39

第 5 章 Android 常用 UI 组件 ························ 40

5.1 使用基于 XML 的布局 ························ 40
5.2 Android 基本组件 ························ 44
5.2.1 Button ························ 44
5.2.2 TextView ························ 47
5.2.3 ImageView ························ 49
5.2.4 EditText ························ 49
5.2.5 CheckBox ························ 50
5.2.6 RadioButton ························ 50
5.3 同步练习一 ························ 50
5.4 Android 容器组件 ························ 50
5.4.1 LinearLayout ························ 50
5.4.2 RelativeLayout ························ 55
5.4.3 FrameLayout ························ 57
5.4.4 ScrollView ························ 60
5.4.5 CoordinatorLayout ························ 63
5.5 同步练习二 ························ 64
5.6 AdapterView ························ 64
5.6.1 AdapterView 入门 ························ 64
5.6.2 Adapter ························ 65
5.6.3 ListView ························ 66
5.6.4 Spinner ························ 77
5.6.5 GridView ························ 83
5.7 同步练习三 ························ 87
5.8 Android 其他常用组件 ························ 87
5.9 同步练习四 ························ 87

第 6 章 样式和主题 ························ 88

6.1 样式入门 ························ 88
6.2 定义样式 ························ 92
6.2.1 定义样式的一般方法 ························ 92
6.2.2 样式定义中的可用属性 ························ 93
6.3 应用样式 ························ 94
6.3.1 将样式应用到某个组件 ························ 95
6.3.2 将样式应用到某个 Activity 或整个 Application ························ 95
6.4 使用 Android 平台已定义的样式和主题 ························ 96
6.5 Android 应用程序的主题样式结构分析 ························ 97
6.6 同步练习 ························ 98

第 7 章 理解和使用 Intent ·· 99

7.1 Intent 应用入门案例 ·· 99
7.2 同步练习一 ·· 103
7.3 细说 Intent ·· 103
7.3.1 Intent 的 action ·· 106
7.3.2 Intent 的 data ·· 106
7.3.3 Intent 的 category ·· 108
7.3.4 Intent 的 extra ··· 109
7.4 Intent 解析 ··· 109
7.5 获得 Activity 返回的结果 ··· 109
7.6 Intent 的综合应用举例 ··· 115
7.6.1 运行效果 ·· 115
7.6.2 程序代码 ·· 117
7.7 同步练习二 ·· 121
7.8 广播消息和广播接收器 ·· 121
7.8.1 发送和接收普通消息 ·· 122
7.8.2 接收 Android 平台的广播消息 ·· 127
7.9 同步练习三 ·· 127

第 8 章 菜单和 Toolbar ··· 128

8.1 菜单 ··· 128
8.2 ActionBar 和 Toolbar ·· 132
8.3 同步练习 ··· 135

第 9 章 动画 ··· 136

9.1 View 动画之补间动画基础 ·· 136
9.1.1 补间动画举例 ·· 136
9.1.2 补间动画类型 ·· 139
9.1.3 使用动画监听器 ··· 142
9.2 View 动画之帧动画 ·· 144
9.3 同步练习 ··· 147

第 10 章 多媒体播放 ··· 148

10.1 使用 MediaPlayer 播放音频 ·· 148
10.1.1 播放简短的音频 ··· 148
10.1.2 使用 MediaPlayer 自制一个音频播放器 ·· 151
10.2 同步练习一 ·· 165
10.3 播放视频 ··· 165
10.4 同步练习二 ·· 167

第 11 章　保存程序数据 … 168

11.1　使用 SharedPreferences 保存程序数据 … 168
11.2　同步练习一 … 172
11.3　设置程序首选项 … 172
11.4　同步练习二 … 180
11.5　在程序目录下存储程序数据 … 181
11.6　同步练习三 … 181
11.7　访问外部存储器 … 181
11.7.1　检查 SD 卡状态 … 182
11.7.2　获得 SD 卡上特定子目录的 File 对象 … 182
11.8　使用 SQLite 数据库保存程序数据 … 183
11.8.1　SQLite 数据库介绍 … 183
11.8.2　在 Android 中使用 SQLite 数据库 … 183

第 12 章　使用后台任务 … 194

12.1　使用 Java 线程执行后台任务 … 194
12.2　同步练习一 … 198
12.3　使用 AsyncTask 执行后台任务 … 198
12.4　使用 Service 完成后台任务 … 203
12.5　同步练习二 … 211

第 13 章　使用网络 … 212

13.1　使用 ConnectivityManager 管理网络状态 … 212
13.2　使用 HttpURLConnection 访问网络 … 214
13.2.1　使用 HttpURLConnection 的 GET 方法获取图片 … 215
13.2.2　使用 HttpURLConnection 的 POST 方法获取图片 … 220
13.3　同步练习一 … 223
13.4　使用 OkHttp 访问网络 … 223
13.4.1　使用 Get 方法进行服务请求 … 223
13.4.2　使用 Post 方法进行服务请求 … 224
13.4.3　设置请求头及提取响应头 … 226
13.4.4　配置 OkHttp 超时 … 226
13.5　OkHttp Get 实现示例 … 227
13.6　OkHttp Post 实现示例 … 231
13.7　同步练习二 … 235
13.8　使用 Multipart 传递请求数据到服务器端程序 … 235
13.9　同步练习三 … 241
13.10　使用 JSON 格式的数据与服务器端通信 … 241

13.10.1 JSON 基础 …………………………………………………………………241

13.10.2 在 JavaScript 中使用 JSON 数据 ……………………………………242

13.10.3 在 Java 中使用 JSON 数据 ……………………………………………242

13.10.4 使用 POST 方法及 JSON 数据格式发送请求 ………………………243

第 1 章

Android 概述

说到智能手机，大家可能马上就能联想到 Android（中文翻译为"安卓"）、iOS 和 Windows 等手机操作系统。确实，在当今，智能手机是如此的流行，以至于上至老年人，下至几岁孩童都在使用智能手机。市场研究公司 IDC 称，在 2018 年，Android 手机销量在全球市场上的占比达到 85.2%。在未来五年内，Android 手机的销量有望达到 2.4% 的复合年均增长率，到 2022 年将达到 14.1 亿部。

可以说 Android 占据了绝对优势。因此，作为应用开发人员，开发基于 Android 的手机应用是一个重要的方向。

1.1 Android 是什么

Android 是如此的流行，那么，Android 到底是什么？这个问题有点抽象，但是，作为即将踏入 Android 开发阵营的读者需要了解这个问题的答案。注意，只需要了解就可以了，即使在阅读这段文字时对有些概念不太理解，这也是正常的，不会影响将来开发出高水平的 Android 应用。

Android 是什么？Android 是一个平台，它包括基础系统、开发工具和完整的文档。Android 平台是一个通用的计算平台，它采用 Linux 为其支撑操作系统，以 Java 作为其开发环境，通过编程实现完整的电话、视频、网络、界面设计等基础功能。Android 平台的体系结构如图 1-1 所示。

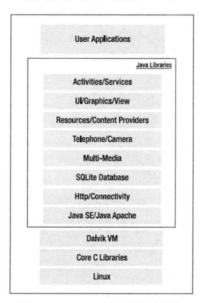

图 1-1　Android 平台的体系结构

我们采用从下往上的方式来介绍Android平台的体系结构的组成部分。

从图1-1可以看到，处在Android平台的底层的是Linux操作系统，在Android的平台体系结构下，Linux提供了基础的支撑功能，包括设备管理、进程管理、文件管理等功能。按理说，Google完全可以开发一套自己的操作系统而不采用Linux操作系统作为Android平台的支撑，可是，Linux是如此成熟，应用又是如此广泛，并且还是开源的系统，为什么不用呢？为什么还要自己开发别人已经做得很好并且可以开源和自由使用的系统呢？这可能是Google使用Linux作为Android平台支撑的原因吧。

在图1-1中，从下往上，可以看到"Core C Libraries"，翻译为中文就是"核心C语言程序库"。为什么是这样呢？如前所述，Linux是一个基础的支撑系统，可是在编写应用程序时如何使用Linux提供的功能呢？Linux为了可以让应用程序使用它提供的功能而专门提供了一系列的C语言函数给应用程序调用，进而应用程序可以充分利用Linux操作系统提供的功能来达到应用程序的业务目标。作为Android应用开发者，除非将来要开发特定功能的应用程序，在一般情况下，读者不会直接使用到这些C语言函数库。当某一天要使用这个"Core C Libraries"开发Android应用程序时，恭喜你，你是Android应用开发的行家了。

图1-1再往上就是"Dalvik VM"，什么是Dalvik VM？它没有那么神秘，Dalvik VM其实就是一个变形的Java虚拟机（Java VM），也就是一个与普通Java虚拟机有一点区别的Java虚拟机而已。可是，Google为什么要实现一个变形的Java虚拟机呢？原因很简单，因为普通的Java虚拟机在普通的桌面计算机上可以运行得很好，可是在小型的设备，如手机上就不一定了，为此，Google实现了一个变形的Java虚拟机，并命名为Dalvik VM。还有一个问题需要回答：为什么Dalvik VM需要出现在这里呢？Android应用开发是基于Java语言进行的，Dalvik VM就像我们开发桌面Java应用程序一样，我们需要一个Java虚拟机来运行开发的Java程序。

图1-1再往上就是一组用一个框框起来的"Java Libraries"，也就是一组Java类库。我们先来问第一个为什么：为什么Java Libraries会出现在这里？我们已经知道，Android应用程序是基于Java语言开发的，为了开发Java应用程序，我们需要使用Java类库来完成某些基本的功能，因此，这里的"Java Libraries"就是在编写Android应用程序时可以使用的一组Java类库。第二个问题：这个Java Libraries中都包含了哪些Java类库？Java Libraries中很多类库读者可能都不认识，也应该不认识，因为这些不认识的类库正是本书要介绍的内容，可是其中的一个读者一定认识：Java SE。是的，这个Java SE就是读者学习Java语言时的那个Java SE，这个类库中包含像String、Integer、File等基础的类库，读者可以像使用Java SE类库中的类一样来使用其他类库中的类。

图1-1再往上就是"User Applications"，也就是用户应用程序，我们将来开发的所有Android应用程序都归属于这里。那么，如何来开发Android应用程序呢？这就是本书的目标。

我们从下往上介绍了Android平台的体系结构，现在，我们再从上往下将以上介绍的内容串起来，然后来看看如何开发一个Android应用程序，以及开发的Android应用程序是如何在Android设备上运行的：采用Java语言来开发一个Android应用程序（也就是一个User Application），在开发应用程序时，可以使用"Java Libraries"中提供的Java类库来实现所要求的功能，当开发完成以后，可以在Dalvik VM上运行应用程序，Dalvik VM会解释Java代码并在Linux操作系统上来执行代码，进而完成Android应用所要完成的业务功能。

1.2 Android 应用程序的组成

我们对 Android 体系有一个初步了解后，作为一个开发者，读者一定想知道 Android 是由哪些部分组成的。如下这些概念对于一个 Android 应用开发初学者来说，有些难于理解，但不要太纠结于这些概念，知道这些概念就好。我们会尽量采用通俗易懂的语言来介绍这些概念，使读者对 Android 程序的组成有一个初步的了解。

任何一个应用程序都会包括如下一些基本内容：应用程序的界面、业务功能的处理、部件之间的数据交互、数据存储。如此而已，Android 应用程序也不例外，只是被赋予了不同的名称而已。具体来说，一个 Android 应用程序包括如下的基本组成部分。

（1）Activity（窗体）。

在 Android 应用程序中，一个界面就是一个 Activity。这个名字有点与众不同，Google 经常会创造一些新名称，不是吗？既然 Activity 就是一个界面，因此，对每个 Activity，在进行设计时，都包括对界面的布局（Android 提供了丰富的 UI 组件来实现绚丽的界面）、对界面组件的点击会进行相应的事件处理等程序设计工作。可以将 Android 的 Activity 界面类比为 Internet 网页的一个页面。

（2）View（窗体组件）。

View 就是构建应用程序界面的基本组件，也就是说，Activity 界面是由一个或多个 View 构成的，例如，Button、Label、Text Field 等都是 View，View 是构建 Activity 的基本元素。

（3）Intent（窗体间或应用之间的通信组件）。

一般来说，一个 Android 应用程序会包括多个界面，用户在进行不同的操作时可能会进行不同界面之间的切换，就像在 Internet 的页面之间，当用户点击不同的页面链接时会进行不同页面之间的切换一样。在 Internet 页面之间的切换是通过链接来完成的，在 Android 的 Activity 之间实现不同的 Activity 界面之间的切换是通过称为 Intent 的对象来完成的。因此，可以这么说，Intent 是 Android 应用程序界面之间及功能部件之间实现信息交互的桥梁。

（4）Content Provider（应用之间数据交互方式）。

Content Provider，也就是内容提供者，是 Android 建议的应用程序之间进行数据交互的方式。举例来说，如果一个应用程序希望将自己的数据提供给其他应用程序使用，则该应用程序需要实现 Content Provider 接口，这样其他的应用程序便可以通过这个接口访问这个程序所提供的数据。一个典型的实现了 Android 的 Content Provider 接口的程序是通讯录程序，任何需要使用通讯录数据的程序都可以通过该接口从电话通讯录程序中获得通讯录数据。

（5）Service（无窗口的在后台默默运行的程序）。

所谓 Service 就是运行于后台的程序。一般来说，Service 程序没有用户界面，它们运行于后台并为运行在前端的程序提供服务。Android 的 Service 程序在运行方式上类似于 Windows Phone 操作系统中的后台进程：它们在安静的运行，并在需要的时候为其他程序提供服务。

（6）广播接收器（信息广播方式）。

广播接收器，即 Broadcast Receiver。Android 平台中的程序在运行时会发生任何可能的事件，某个应用程序在运行时可能会将它的事件广播出来，其他的程序可以监听这样的事件，并对发生的事件进行必要的处理。举例来说，Android 的电池电量监视程序（这是 Android 的一个 Service）在随时监视着电池的电量，当电池的电量低于某个门槛值时，该程序会广播一个消息，而其他的应用程序可以监听这个消息，并针对这个事件做出必要的处理，例如，一个正在进行高耗电运算的程序监听到这个消息时，应该停止进行高耗电的运算，以便减少对电量的消耗。

（7）AndroidManifest.xml 文件（应用程序描述文件）。

AndroidManifest.xml 文件是 Android 应用程序的配置文件：它将构成 Android 应用程序的各个组件有效地装配起来从而构成一个完整的 Android 应用程序。每个应用程序一定会包含一个且只能包含一个配置文件。

1.3 Android 的发展历史

手机的常用操作系统有：Symbian OS、Microsoft Windows Mobile、Mobile Linux、iOS 等，这其中没有任何一个操作系统是事实上的标准，也没有任何一个操作系统是开源的。为此，Google 创立了 Android 移动平台。

2007 年，Google 牵头建立了开放手机联盟（Open Handset Alliance），到 2009 年，这个联盟成员包括 Sprint Nextel、T-Mobile、Motorola、Samsung、Sony Ericsson、Toshiba、Vodafone、Google、Intel、Texas Instruments 等 IT 巨头，到 2011 年，成员已近 80 家，Android 已经成为移动设备事实上的行业标准。到 2022 年 1 月，Android 平台的版本已从 1.0 发展到了 12.0。Android 是进化得如此之快，因此，读者在学习 Android 开发时，必须要学会使用 Android 的在线帮助文档。

1.4 Android 应用开发环境概述

Android 平台采用 Java 语言作为应用程序开发语言。Android 开发环境包括如下基本内容：Java 基本包、Android 基础组件、Android UI 组件、Android 服务组件、Android 电话和媒体服务组件、Android 仿真器（Android Virtual Device，AVD）、Android 调试器等。

笔者将在后续章节中，对这些内容进行详细介绍。首先从建立 Android 开发环境开始讲解。

第 2 章

建立 Android 应用开发环境

由于 Android 应用是基于 Java 语言进行开发的,因此,首先需要安装 Java SE 环境,笔者建议安装 Java JDK 8 或以上版本。基于 Java JDK,Android 平台提供了两种建立开发环境的方式。

(1)使用基于 Eclipse 和 ADT 的应用开发环境;
(2)使用基于 Android Studio 的应用开发环境。

由于第一种方式使用 Eclipse IDE 作为开发环境,因此,对于熟悉 Eclipse 的开发者来说使用第一种方式会更为合适,第二种方式,也就是基于 Android Studio 的方式,是 Android 提供的非 Eclipse IDE 的开发环境。

由于 Google 将逐渐取消对 Eclipse 开发环境的支持,因此,我们将使用 Android Studio 作为本书的开发环境。为了测试 Android 应用程序,我们可以使用 Android SDK 自带的模拟器,也可以使用第三方的模拟器如 Genny Motion 来测试应用程序。在本书中,我们使用 Android SDK 自带的模拟器来测试程序。

2.1 下载和安装 Android Studio

在下载和安装 Android Studio 之前,需要先下载和安装 Java SDK。从 Oracle 网站下载并安装就可以了。在安装 Java JDK 之后,从 Android 的开发者网站下载 Android Studio,运行下载得到的文件即可安装 Android Studio 开发环境。

2.2 开发第一个 Android 应用程序

2.2.1 创建 Hello World 程序工程

我们已经建立了 Android 应用开发环境,现在就可以基于这个环境开发 Android 应用程序了。启动 Android Studio,即可显示如图 2-1 所示的界面。

图 2-1　Android Studio 启动界面

按照提示安装 Android Studio 开发环境即可。安装完毕后，运行 Android Studio 开发环境，单击第一个功能选项"Start a new Android Studio project"，即可创建一个新的 Android 应用程序工程，如图 2-2 所示。

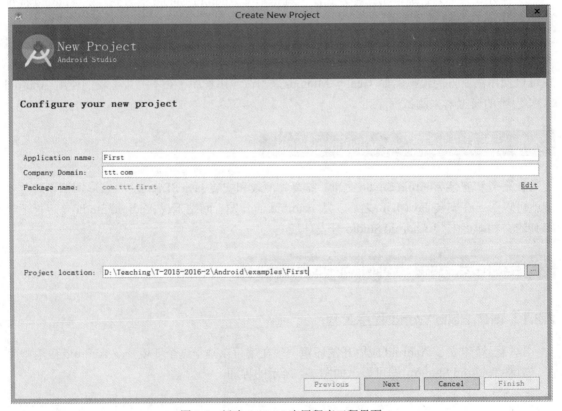

图 2-2　新建 Android 应用程序工程界面

在如图 2-2 所示的界面中，在"Application name"中，输入应用程序的名字，在"Company Domain"中输入公司的域名（可以是假想的域名），在"Project location"中输入应用程序文件存放的位置，然后单击"Next"按钮，即可显示如图 2-3 所示的界面。

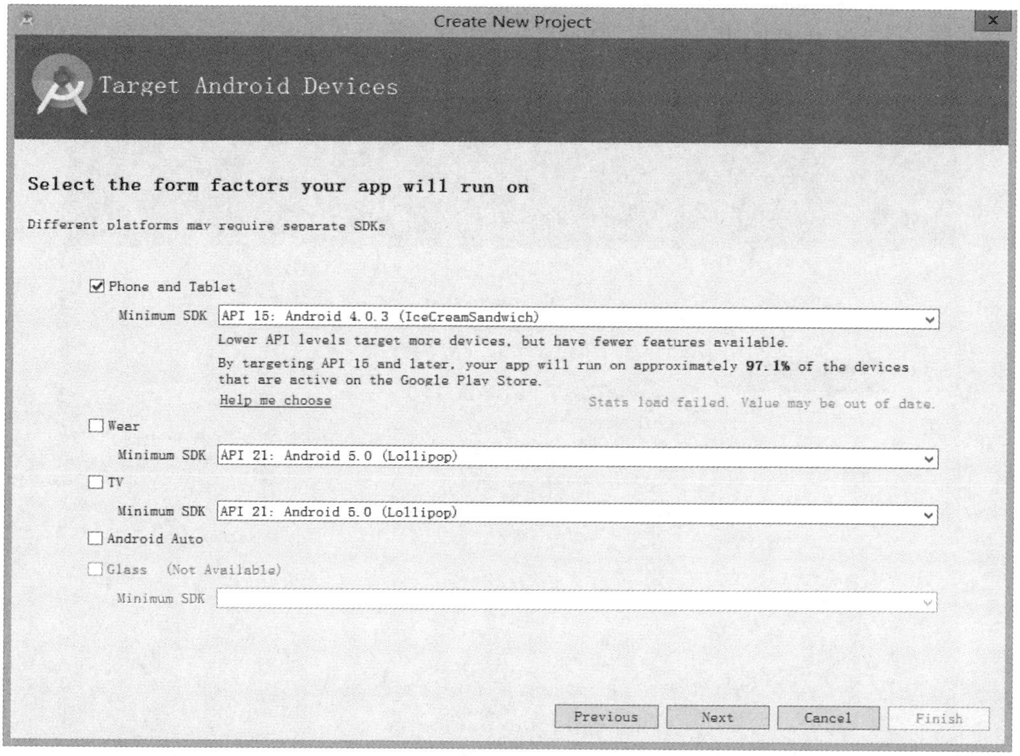

第 2 章 建立 Android 应用开发环境

图 2-3 应用程序运行目标机设置

在如图 2-3 所示的界面中，选择应用程序运行的目标机，直接单击 "Next" 按钮，即可显示如图 2-4 所示的界面。

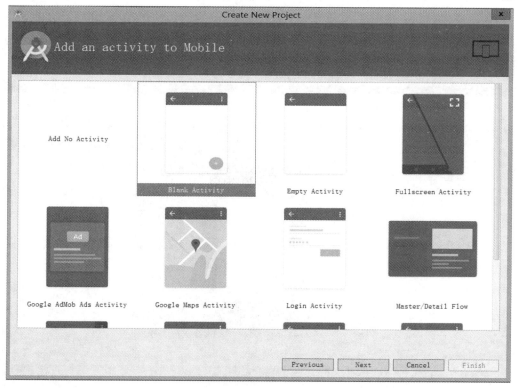

图 2-4 选择应用界面框架

在如图 2-4 所示的界面中,选择"Blank Activity",单击"Next"按钮,即可显示如图 2-5 所示的界面。

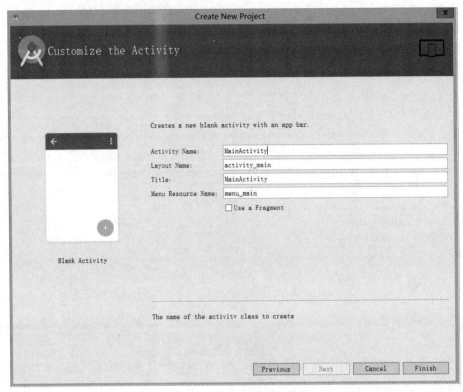

图 2-5 Android 工程概览

在如图 2-5 所示的界面中,单击"Finish"按钮,Android Studio 将创建一个新的 Android App 工程,Android Studio 创建工程文件需要一些时间,创建成功后,即可显示如图 2-6 所示的界面。

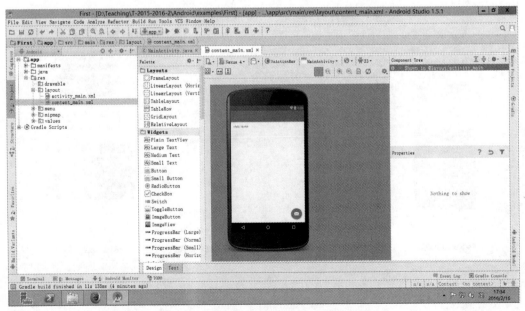

图 2-6 成功创建 Android App 工程的界面

至此，Android Studio 已经成功创建了名称为"First"的 Android App 工程。现在我们可以运行这个最简单的程序了。

2.2.2 运行 Hello World 程序

测试 Android 应用程序，既可以直接在 Android 真机上测试，也可以在 Android SDK 自带的模拟器上测试。我们首先在 Android 模拟器上测试，包括 3 个步骤：第一，下载需要的 Android 工具和版本；第二，配置一个模拟器；第三，在模拟器上运行程序。

1. 下载需要的 Android 工具和版本

为了能够在模拟器上测试程序，我们需要首先从 Android 站点下载必需的 Android 平台及其相应的支持工具。为此，进入在安装 Android Studio 时选择的 Android SDK 的安装目录，会看到与如图 2-7 所示类似的目录内容。

图 2-7　Android SDK 目录内容

运行"SDK Manager.exe"程序，即可显示如图 2-8 所示的界面。

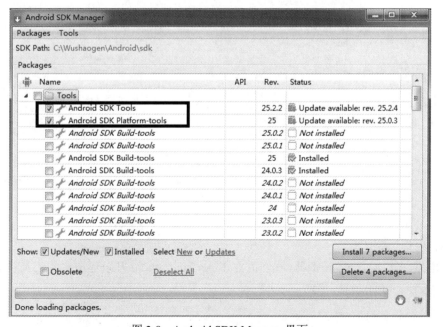

图 2-8　Android SDK Manager 界面

在"Tools"类别下,确保最新的"Android SDK Tools"和"Android SDK Platform-tools"被选中。在 Android 系统的版本下,确保如图 2-9 所示的常用的 Android 版本被选中。

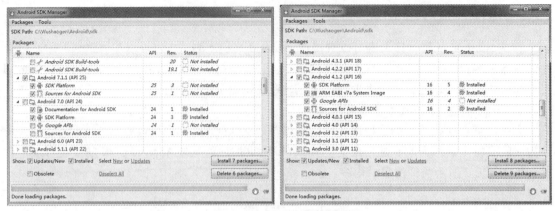

图 2-9 常用的 Android 版本

然后,在"Extras"类别下,选择如图 2-10 所示的基础组件。

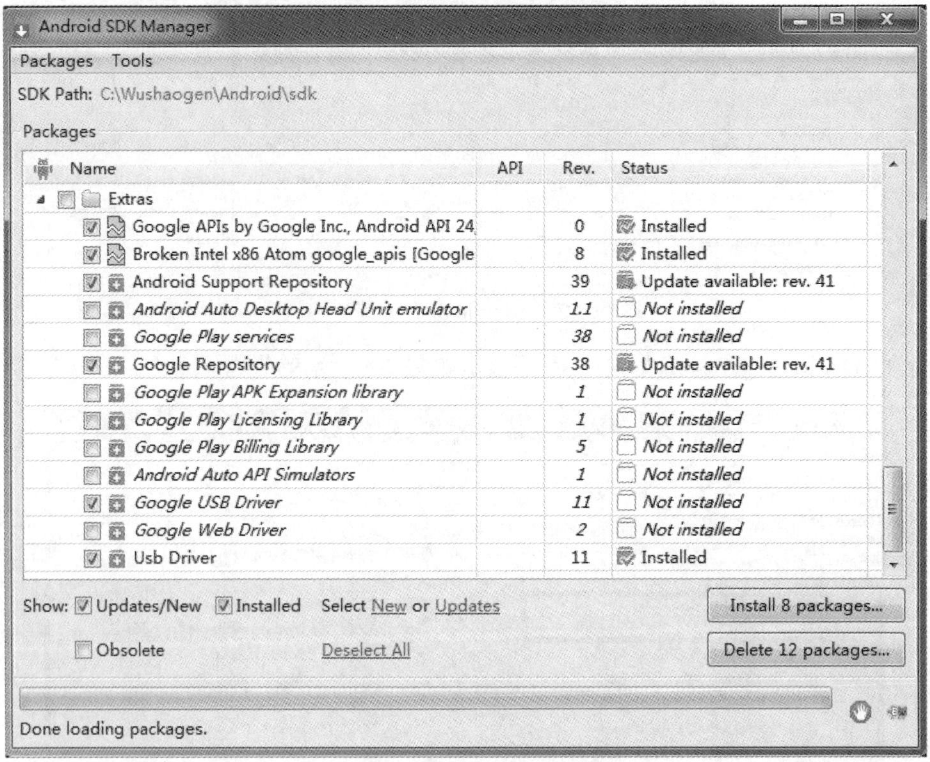

图 2-10 "Extras"基础组件

最后,单击"Install"按钮,安装所有被选中的组件。根据网络速度,这会需要一些时间。

2．配置一个模拟器

为了配置一个 Android 模拟器,我们需要在 Android Studio 开发环境中的工具栏单击"AVD Manager"按钮,如图 2-11 所示。

第 2 章 建立 Android 应用开发环境

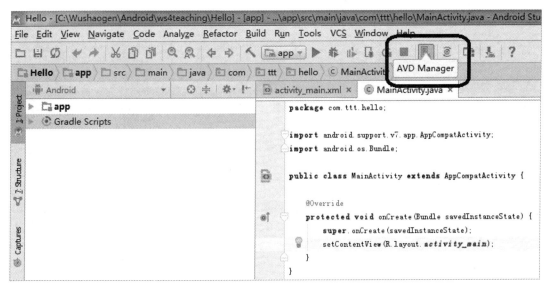

图 2-11 单击"AVD Manager"按钮

在"AVD Manager"显示的界面单击"Create Virtual Device"按钮，即可显示如图 2-12 所示的界面。

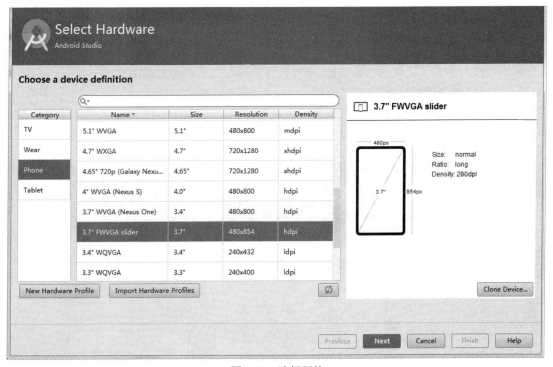

图 2-12 选择硬件

在如图 2-12 所示的界面中，选择"Phone"及"3.7" FWVGA slider"，然后单击"Next"按钮，即可显示如图 2-13 所示的界面，选择"Other Images"和"Jelly Bean"的 Android 影像。注意，这里所选择的影像必须与在第一步中选择的 Android 系统版本相匹配，例如，因为在如图 2-9 所示的界面中选择了 Android 4.1.2（API 16）及 ARM EABI v7a 的系统影像，因此，我们在这一步就可以使用该版本的影像来建立模拟器。当然，如果之前没有选择需要的影像，也可以在如图 2-12 所示的界面进行下载。

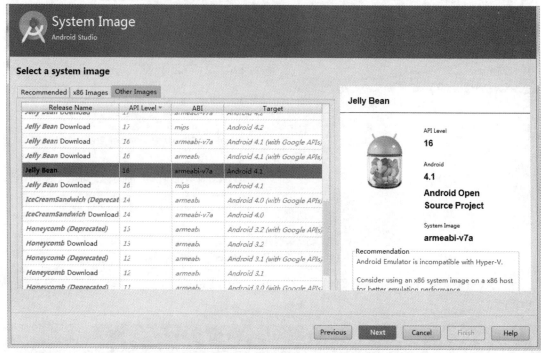

图 2-13 选择指定的 Android 影像

单击"Next"按钮，然后单击"Finish"按钮，即可完成模拟器的配置。

3．在模拟器上运行程序

为了能够在模拟器上运行程序，需要先启动模拟器。在如图 2-14 所示的界面中单击三角形按钮，即可启动刚才所配置的模拟器。

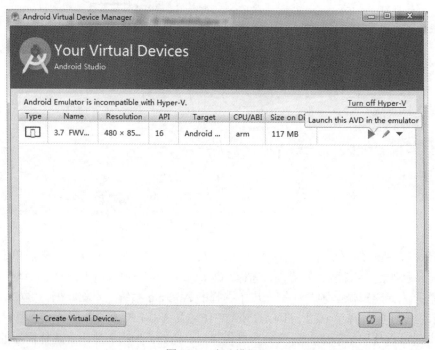

图 2-14 启动模拟器

模拟器启动后，即可显示如图 2-15 所示的手机界面。

图 2-15　模拟器启动后的手机界面

现在，可以测试"Hello World"程序了。在 Android Studio 界面单击如图 2-16 所示的启动应用程序按钮。

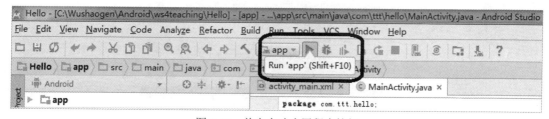

图 2-16　单击启动应用程序按钮

在显示的界面中单击"OK"按钮，即可运行"Hello World"程序，如图 2-17 所示。

图 2-17　Hello World 程序

2.3　Android 应用程序的结构

由于 Android 应用环境的不同，Android 应用程序的规模也相差很大，但是，Android 应用程序的结构是相似的。典型的 Android 应用程序的结构如图 2-18 所示。

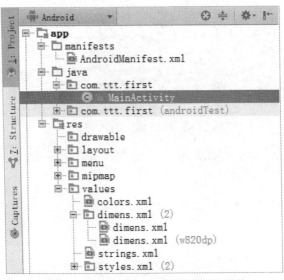

图 2-18　典型的 Android 应用程序的结构

从上往下，我们逐个对 Android 应用程序的工程结构进行介绍。

1. manifests 目录下的 AndroidManifest.xml 文件

这个文件是 Android 应用程序的配置文件，它类似于 Java EE 程序的 web.xml 文件。它描述了 Android 应用程序的所有必要的信息，包括一个 Android 程序是由哪些 Activity 组成的、哪个是入口 Activity、运行该 Android 程序需要的 Android 平台的最低版本等。

2. java 目录

java 目录是应用程序的 Java 源代码程序包和源代码所在的位置，例如，第一个 Hello World 程序的 com.ttt.first 包及源代码文件 MainActivity.java 都保存在这个目录下。

3. res 目录

这个目录下包括几个子目录，典型的包括 drawable、layout、values 等，这个目录与 java 目录一样重要，是经常需要操作的地方。这个目录用于存放 Android 应用程序的称为"资源"的东西。我们会在后续章节对这个目录下的资源类型及其格式进行详细介绍。

至此，我们已经对 Android 平台及 Android 应用程序开发环境进行了初步的介绍，同时，我们还开发了第一个 Android 应用程序。接下来，我们将对 Android 应用程序进行剖析，带领读者逐步进入 Android 的精彩世界。

从工程的角度来说，一个 Android 工程的 Project 视图如图 2-19 所示（单击 Android Studio 界面左上角的下拉按钮，在弹出的下拉列表中选择"Project"）。

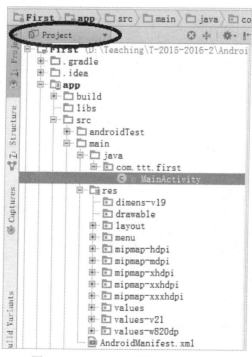

图 2-19 Android 工程的 Project 视图

对于该视图中包含的文件及结构，暂时不用关心它们。

2.4 同步练习

建立 Android 开发环境，然后编写第一个 Hello Android 程序，运行它，并观察程序的结构。同时，熟悉 Android Studio 开发环境的各个功能的使用方法。

第 3 章

剖析 Android 应用程序

在第 2 章中,我们对 Android 应用程序的工程结构进行了简单的介绍,并指出 AndroidManifest.xml 文件是 Android 应用程序非常重要的配置文件,在 Android 应用程序运行时,Android 平台将首先读取这个文件并进行分析,然后再启动特定的 Activity 来运行。因此,我们就从这个文件开始对 Android 应用程序进行剖析。

3.1 AndroidManifest.xml

双击打开第 2 章介绍的 Hello World 工程例子中的 AndroidManifest.xml 文件,将看到如下内容,这是一个 XML 文件:

```xml
<?xml version="1.0" encoding="utf-8"?>
<manifest xmlns:android="http://schemas.android.com/apk/res/android"
package="com.ttt.first">

<application
      android:allowBackup="true"
      android:icon="@mipmap/ic_launcher"
      android:label="@string/app_name"
      android:supportsRtl="true"
      android:theme="@style/AppTheme">
      <activity
         android:name=".MainActivity"
         android:label="@string/app_name"
         android:theme="@style/AppTheme.NoActionBar">
         <intent-filter>
            <action android:name="android.intent.action.MAIN" />
            <category android:name="android.intent.category.LAUNCHER" />
         </intent-filter>
      </activity>
</application>
</manifest>
```

如前所述,AndroidManifest.xml 文件描述了 Android 应用程序的基本信息,其中的

```xml
<?xml version="1.0" encoding="utf-8"?>
```

说明 AndroidManifest.xml 是一个 XML 文件。<manifest>标签如下:

```xml
<manifest xmlns:android="http://schemas.android.com/apk/res/android"
   package="com.ttt.first"
```

<manifest>标签中首先通过 xmlns:android 属性定义了 android 前缀的名字空间,以避免名字空间的冲突,然后,通过 package 属性定义了应用程序的包名,注意,每个 Android 应用程序都有一个唯一的包名,这个唯一的包名就是通过<manifest>标签的 package 属性来定义的,这个包名也是我们在创建 Android 工程时定义的包名,我们称这个包为"应用的包"。接下来:

```
<application
    android:allowBackup="true"
    android:icon="@mipmap/ic_launcher"
    android:label="@string/app_name"
    android:theme="@style/AppTheme" >
```

<application>标签定义了一个 Android 应用程序,其中的 android:icon 和 android:label 分别定义了应用程序安装后显示在 Android 手机的应用程序管理中的图标和名称,在 Android 手机上,可以通过 Settings→Apps 进入应用程序管理界面,如图 3-1 所示。

图 3-1　Android 的应用程序管理界面

图 3-1 中框住的部分,分别为应用程序图标和应用程序名称。有的读者可能会问:@mipmap/ic_launcher 和@string/app_name 又是在哪里定义的呢?Android 应用程序中用到的图标、字符串常量等,都称为 Android 应用程序资源,它们是在工程目录下的 res 目录下定义的,如图 3-2 所示。

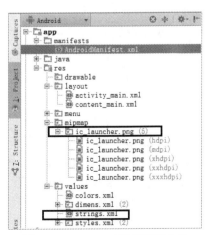

图 3-2　Android 工程的资源定义

如果打开 drawable 目录下的 ic_launcher.png 文件，会显示一个小机器人的图标，这个图标就是显示在应用列表中的应用程序图标。同时，如果打开 values 目录下的 strings.xml 文件，将看到如下的内容：

```
<resources>
    <string name="app_name">First</string>
    <string name="action_settings">Settings</string>
</resources>
```

其中的 app_name 就对应于@string/app_name 指定的常量字符串。

我们继续观察 AndroidManifest.xml 文件的内容，接下来的内容是：

```
    <activity
        android:name=".MainActivity"
        android:label="@string/app_name" >
```

这段 XML 代码定义了应用的一个 Activity，也就是 Android 应用程序的一个界面。Android 应用程序的一个 Activity 相当于网站的一个页面，这个 Activity 的主类由 android:name 定义，<manifest>标签的 package 属性的值加上 android:name 属性的值就构成了 Activity 的主程序类，因此，这个 Activity 的主类就是 com.ttt.helloworld.MainActivity。android:label 定义了 Activity 在手机上显示时，显示在屏幕上方的标签，如图 3-3 所示。

图 3-3 Activity 显示的标签

图 3-3 中框住的部分，就是这个 Activity 的 android:label 定义的值。

我们继续观察 AndroidManifest.xml 文件的内容，如下所示：

```
    <intent-filter>
        <action android:name="android.intent.action.MAIN" />
        <category android:name="android.intent.category.LAUNCHER" />
```

这段 XML 代码定义了"入口 Activity"，现在来解释如何定义入口 Activity。与其他的应用程序类似，Android 应用程序也有一个称为入口 Activity 的 Activity 程序，该 Activity 程序是 Android 应用程序的入口：当启动 Android 应用程序时，将首先运行该入口 Activity。它类似于 C 语言程序的 main()函数，网站程序的 index.html，Java 桌面程序的 main()等。那么，如何来定义一个 Activity 为入口 Activity 呢？通过<intent-filter>标签来定义：在<intent-filter>标签中，通过定义 android:name 属性为 android.intent.action.MAIN 的<action>标签和 android:name 属性为 android.intent.category.LAUNCHER 的<category>标签来定义一个 Activity 为入口 Activity。

至此，我们对 AndroidManifest.xml 文件进行了详细的描述，接下来将对 MainActivity.java 程序代码进行解析。

3.2　MainActivity.java——Activity 介绍

在启动 Android 应用程序时，Android 平台将首先读取 AndroidManifest.xml 文件，从中获得入口 Activity 的相关信息，并启动入口 Activity 运行程序。在 Hello World 程序中，由于入口 Activity 是 MainActivity.java 程序，因此，Android 平台将首先启动 MainActivity 程序运行，进而显示程序界面。MainActivity.java 程序代码如下：

```java
package com.ttt.first;

import android.os.Bundle;
import android.support.v7.app.AppCompatActivity;
import android.view.View;

public class MainActivity extends AppCompatActivity {

    @Override
    protected void onCreate(Bundle savedInstanceState) {
        super.onCreate(savedInstanceState);
        setContentView(R.layout.activity_main);
    }
}
```

这段代码很简单。当 Android 平台启动 MainActivity 运行时，系统将首先调用 onCreate() 方法，注意，Android 要求，必须首先调用父类的 onCreate() 方法，然后才可以做一些自己的初始化工作。在 MainActivity.java 程序的 onCreate() 方法中，在调用了父类的 onCreate() 方法之后，调用 setContentView(R.layout.activity_main) 来显示 Activity 的主界面。

在 Android 程序中，有两种构建应用程序界面的方式：①通过布局文件来构建界面；②通过程序代码来构建界面。这里，我们采用布局文件的方式来构建程序界面，这也是常用的构建 Android 程序界面的方法。注意代码中的

```
setContentView(R.layout.activity_main)
```

其中，参数 R.layout.activity_main 就是界面布局文件，它对应工程 res/layout 目录下的 activity_main.xml 文件，如图 3-4 所示。

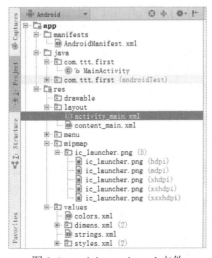

图 3-4　activity_main.xml 文件

读者可以这样理解 R.layout.activity_main：R 代表 res，layout 代表 res 下的 layout，activity_main 代表 res/layout 目录下的 activity_main.xml 资源文件。打开 activity_main.xml，文件内容如下：

```xml
<?xml version="1.0" encoding="utf-8"?>
<LinearLayout xmlns:android="http://schemas.android.com/apk/res/android"
    android:layout_width="match_parent"
    android:layout_height="match_parent">

    <TextView
        android:layout_width="wrap_content"
        android:layout_height="wrap_content"
        android:text="Hello World!" />

</LinearLayout>
```

这是 Android 程序的界面布局资源（Layout Resource）文件。这个布局资源表示：采用 LinearLayout 布局管理器对组件的布局进行管理。同时，这个布局管理器在水平方向将占满整个屏幕，代码为 android:layout_width="match_parent"，在垂直方向也将占满整个屏幕，代码为 android:layout_height="match_parent"。在这个布局管理器中，包含一个称为 Text View 的可视组件，这个组件可用于显示一行文本信息，在水平方向上，这个组件占用的空间只需要足够显示其内部的文字即可。这也就是 android:layout_height="wrap_content" 和 android:layout_width="wrap_content" 的含义。

至此，我们可以很容易地理解下面的语句：

```
setContentView(R.layout.activity_main)
```

这条语句的作用就是：将资源 R.layout.activity_main 所指定的界面显示在 MainActivity 的主窗口中。

请读者根据前面所讲述的内容试着做一道练习：将显示在屏幕上的"Hello World"改为"你好，世界！"

做完这个练习，大家可以一起再做一道练习：在界面上显示的"你好，世界！"的下面再显示一句话"Android 开发入门了！"

在 Android 应用程序中，我们可以在任何需要的地方直接使用字符串常量，就像我们在 TextView 组件中使用的那样。但是，在 Android 中不建议这么做，Android 建议：将字符串常量放置到 res/values/strings.xml 资源文件中进行集中管理。为此，我们修改 res/layout/activity_main.xml 文件为如下内容（读者课堂练习的结果）：

```xml
<?xml version="1.0" encoding="utf-8"?>
<LinearLayout xmlns:android="http://schemas.android.com/apk/res/android"
    android:layout_width="match_parent"
    android:layout_height="match_parent"
    android:orientation="vertical">
    <TextView
        android:id="@+id/tv01"                    //为这个 TextView 赋予一个 ID：tv01
        android:layout_width="wrap_content"
        android:layout_height="wrap_content"
        android:text="@string/hello_world" />

    <TextView
```

```xml
        android:layout_width="wrap_content"
        android:layout_height="wrap_content"
        android:text="@string/my_message" />


</LinearLayout>
```

同时，需要在 res/values/strings.xml 文件中增加对 hello_world 字符串资源的定义，内容如下：

```xml
<?xml version="1.0" encoding="utf-8"?>
<resources>
    <string name="app_name">HelloWorld</string>
    <string name="hello_world">Hello World!</string>          //添加了一条语句
    <string name="my_message">Android开发入门了！</string>    //添加了一条语句
</resources>
```

这种将程序资源与程序代码分离的方式能更好地实现对程序的后期维护，例如，如果我们要使程序界面显示德文，只需要更改资源文件就可以了，而不需要修改程序源代码。Android 程序会用到各种各样的资源，本书将在后续章节中详细介绍关于这些资源的定义和使用方法。从上面的介绍，可得出这样一个结论：

Android 应用程序 = Java 程序代码文件 + 资源文件 + AndroidManifest.xml 文件

粗略来讲这是对的！Android 资源在 Android 程序中扮演着如此重要的角色，下面我们首先对 Android 的常用资源类型进行简单的介绍。

3.3 Android 程序资源

Android 程序资源在 Android 应用程序中起着十分重要的作用：Android 程序资源可以是一个文件，如布局资源，也可以是一个值，如字符串常量定义。将应用程序的资源与应用程序代码分离的好处是：可以直接改变资源的值，而不用修改或编译应用程序代码本身。

在 Android 应用程序中，会用到各种各样的资源，包括字符串资源、图片资源、界面布局资源、动画资源等，下面我们对常用的资源类型的定义和使用进行简单的介绍。

3.3.1 字符串资源

在程序编码实践中，我们经常会用到大量的字符串常量。为此，Android 建议，将字符串常量统一定义到一个或多个 XML 资源文件中，例如，在 Hello World 工程中，我们将用到的字符串常量统一定义到 res/values/strings.xml 文件中。这里需要特别强调：用于定义字符串常量的 XML 文件必须放置到 res/values 工程目录下，而文件名则可以根据需要自行定义。例如，我们可以将 strings.xml 改名为 constants.xml，程序照样可以正常运行。当然，我们根据程序需要，可以将字符串常量定义到多个文件中。例如，我们将原先保存在 strings.xml 文件中的内容拆分到两个文件 constants.xml 和 another.xml 中，只要这两个文件都放置在 res/values 工程目录下，则程序仍然可以正常运行。文件内容如下。

constants.xml：

```xml
<?xml version="1.0" encoding="utf-8"?>
<resources>
    <string name="app_name">HelloWorld</string>
    <string name="hello_world">Hello World!</string>
```

</resources>

　　another.xml：

```
<?xml version="1.0" encoding="utf-8"?>
<resources>
    <string name="my_message">Android开发入门了！</string>
</resources>
```

　　运行修改后的程序，将得到与前面完全一样的结果。

　　在 Android 工程结构中，选择 Packages 视图，在应用程序的包目录下，有一个特殊的 R 文件，如图 3-5 所示。

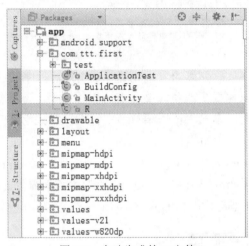

图 3-5　自动生成的 R 文件

　　打开这个文件，内容如下（为了简洁，我们剔除了部分内容和其中的注解）：

```
package com.ttt.helloworld;

public final class R {
    ……
    public static final class string {
        public static final int abc_action_bar_home_description=0x7f060000;
        public static final int abc_action_bar_home_description_format=0x7f060001;
        public static final int abc_action_bar_home_subtitle_description_format=0x7f060002;
        public static final int abc_action_bar_up_description=0x7f060003;
        public static final int abc_action_menu_overflow_description=0x7f060004;
        public static final int abc_action_mode_done=0x7f060005;
        public static final int abc_activity_chooser_view_see_all=0x7f060006;
        public static final int abc_activity_chooser_view_choose_application=0x7f060007;
        public static final int abc_capital_off=0x7f060008;
        public static final int abc_capital_on=0x7f060009;
        public static final int abc_search_hint=0x7f06000a;
        public static final int abc_searchview_description_clear=0x7f06000b;
        public static final int abc_searchview_description_query=0x7f06000c;
        public static final int abc_searchview_description_search=0x7f06000d;
        public static final int abc_searchview_description_submit=0x7f06000e;
        public static final int abc_searchview_description_voice=0x7f06000f;
        public static final int abc_shareactionprovider_share_with=0x7f060010;
        public static final int abc_shareactionprovider_share_with_application=0x7f060011;
        public static final int abc_toolbar_collapse_description=0x7f060012;
```

```
        public static final int action_settings=0x7f050002;      //注意这几句
        public static final int app_name=0x7f050000;
        public static final int hello_world=0x7f050001;
        public static final int my_message=0x7f050003;
        public static final int appbar_scrolling_view_behavior=0x7f060016;
        public static final int character_counter_pattern=0x7f060017;
        public static final int status_bar_notification_info_overflow=0x7f060013;
    }
……
}
```

这个文件的内容是 ADT 自动生成的。在这个称为 R 的类中，包含几个 static 的 final 类，注意其中的名称为 string 的静态类，内容如下：

```
public static final class string {
    public static final int app_name=0x7f050000;
    public static final int hello_world=0x7f050001;
    public static final int my_message=0x7f050003;
}
```

我们观察一下其中的几个 static final 的常数：app_name、hello_world、my_message。这些常数就代表了在 res/values 工程目录下的 strings.xml 文件中定义的字符串内容，只是这里使用了整数类型的编号来代表字符串常量而已。在 Android 中会给每个资源分配一个唯一的整数编号，通过这个编号就可以访问到特定的资源。在 Java 程序代码中，可以使用语句：

```
R.string.app_name
```

来访问由 app_name 定义的字符串；而在资源配置 XML 文件中，如 AndroidManifest.xml 文件中，则使用语句：

```
@string/app_name
```

来访问由 app_name 定义的字符串。

3.3.2 布局资源

在 Android 中，除了可以使用 Java 代码直接构建程序界面，更常用的方式是通过 XML 文件来布局界面，通过 XML 方式来构建程序界面的文件称为布局资源文件，例如，在 Hello World 程序中，res/layout/activity_main.xml 就是一个布局资源文件，内容如下：

```xml
<?xml version="1.0" encoding="utf-8"?>
<LinearLayout xmlns:android="http://schemas.android.com/apk/res/android"
    android:layout_width="match_parent"
    android:layout_height="match_parent">

    <TextView
        android:layout_width="wrap_content"
        android:layout_height="wrap_content"
        android:text="Hello World!" />

</LinearLayout>
```

创建了这个布局资源后，可以通过程序代码直接将该界面显示在 Activity 中，就像我们在 MainActivity.java 程序中操作的那样，内容如下：

```
public class MainActivity extends Activity {
```

```
    @Override
    protected void onCreate(Bundle savedInstanceState) {
        super.onCreate(savedInstanceState);
        setContentView(R.layout.activity_main);    //显示布局界面
    }
    ……
}
```

Android 规定，所有的布局资源文件必须放置在 res/layout 工程目录下。因此，如果有多个界面需要显示，则在 res/layout 工程目录下会有多个这样的布局资源文件。注意代码中的语句：

```
setContentView(R.layout.activity_main);
```

在这里，我们通过如下语句：

```
R.layout.activity_main
```

来引用布局资源。为什么可以这样呢？打开 R 文件就明白了，内容如下：

```
public final class R {
    ……
    public static final class layout {
        public static final int activity_main=0x7f030000;
    }
    ……
}
```

与引用字符串资源一样，在 Java 程序代码中，我们可以通过 R.layout.activity_main 来引用特定的布局资源。

现在对程序的功能稍做改进：我们希望把在第二个 TextView 中显示的字符串 "Android 开发入门了！"改为"欢迎进入 Android 阵营！"，该如何做呢？也就是说，我们需要在程序代码中获得第二个 TextView 的引用，然后通过 TextView 对象提供的函数方法修改在其中显示的字符串。为了能够在 Java 代码中获得布局中某个对象的引用，我们需要对所引用的组件赋予一个 ID。这就是下一节要介绍的内容。

3.3.3 ID 资源

就像我们给字符串资源、布局资源定义一个唯一的标识符一样，我们也可以给布局资源中的各个组件分配一个唯一的 ID。在给布局资源中的组件分配一个唯一的 ID 后，我们就可以在程序代码中引用这些组件了。为此，需要在布局资源中，为每个需要指定 ID 的组件加上 android:id 属性。例如，在 res/layout/activity_main.xml 的布局文件中，我们可以给第二个 TextView 赋予一个 ID，内容如下：

```
<LinearLayout xmlns:android="http://schemas.android.com/apk/res/android"
    android:layout_width="match_parent"
    android:layout_height="match_parent"
    android:orientation="vertical">

    <TextView
        android:id="@+id/tv01"                    //为这个 TextView 赋予一个 ID：tv01
        android:layout_width="wrap_content"
        android:layout_height="wrap_content"
        android:text="@string/hello_world" />
```

```xml
    <TextView
        android:id="@+id/tv02"                    //为这个TextView赋予一个ID：tv02
        android:layout_width="wrap_content"
        android:layout_height="wrap_content"
        android:text="@string/my_message" />

</LinearLayout>
```

在上面的代码中，我们给第一个 TextView 组件分配名为 tv01 的 ID；给第二个 TextView 组件分配名为 tv02 的 ID。在程序代码中，我们可以通过如下语句：

```
TextView tv = (TextView)this.findViewById(R.id.tv02);
```

来获得对 ID 为 tv02 的组件的引用，然后通过如下语句：

```
tv.setText("欢迎进入 Android 阵营！");
```

来修改显示在组件中的内容。修改后的完整的 MainActivity.java 代码如下：

```java
package com.ttt.helloworld;

import android.app.Activity;
import android.os.Bundle;
import android.widget.TextView;


public class MainActivity extends AppCompatActivity {

    @Override
    protected void onCreate(Bundle savedInstanceState) {
        super.onCreate(savedInstanceState);
        setContentView(R.layout.activity_main);
        ……

        TextView tv = (TextView)this.findViewById(R.id.tv02);
        tv.setText("欢迎进入 Android 阵营！");
    }
        ……
}
```

程序修改后的结果如图 3-6 所示。

图 3-6　通过代码修改显示在组件中的信息

这正是我们期望的结果。

如前所述,在 Android 开发实践中,我们一般都是将字符串资源统一定义在 res/values 目录下,为此,我们在 res/values/strings.xml 文件中添加一行代码,内容如下:

```xml
<?xml version="1.0" encoding="utf-8"?>
<resources>

    <string name="app_name">HelloWorld</string>
    <string name="hello_world">Hello World!</string>
    <string name="my_message">Android开发入门了!</string>
    <string name="message_2">欢迎进入Android阵营!</string>        //增加了这一行

</resources>
```

同时修改 MainActivity.java 的 onCreate()方法为如下内容:

```java
public class MainActivity extends AppCompatActivity {

    @Override
    protected void onCreate(Bundle savedInstanceState) {
        super.onCreate(savedInstanceState);
        setContentView(R.layout.activity_main);
        ……

        TextView tv = (TextView)this.findViewById(R.id.tv02);
        //tv.setText("欢迎进入Android阵营!");
        tv.setText(R.string.message_2);        //将注释掉的代码替换为这一句
    }
```

运行程序,将得到同样的结果,但是,这种编码方式更符合 Android 的要求。如果此时打开 R 文件,会发现有如下变化:

```java
public final class R {
    ……
    public static final class id {
        ……
        public static final int tv01=0x7f080000;            //ID 资源
        public static final int tv02=0x7f080001;
        ……
    }
    ……
    public static final class string {
        ……
        public static final int app_name=0x7f050000;
        public static final int hello_world=0x7f050001;
        public static final int message_2=0x7f050004;      //新的 String 资源
        public static final int my_message=0x7f050003;
        ……
    }
    ……
}
```

我们所定义的新的资源均正确地反映在 R 文件中。

3.3.4 图片资源

为了实现程序的可用性和美观性,在程序中一般都会用到图片来装饰界面。

Android 将图片也定义为资源,它将每个图片资源作为文件放置在 res/mipmap(或 res/drawable) 工程目录下。有的读者可能会问,为什么在 Android 的工程目录下没有 res/mipmap 目录呢?其实, res/mipmap-ldpi、res/mipmap-mdpi 和 res/mipmap-hdpi 都是图片资源目录。为了弄明白这一点,有一个事实需要读者先了解清楚:Android 是一个支持国际化和多设备类型的平台,为了支持国际化和多设备类型,Android 将代码与资源进行分离管理。那么,如何使一个资源能同时满足多种语言及设备类型呢?Android 通过采用资源目录后缀的方式来达到这个目标。例如,对于图片资源,为了满足不同的屏幕分辨率和屏幕大小,我们可以对不同的屏幕分辨率和屏幕大小创建不同的资源,对于高分辨率的设备,则会自动从 res/mipmap-hdpi 目录下获得图片资源;对于低分辨率的设备,则会自动从 res/mipmap-ldpi 目录下获得图片资源;对于中等分辨率的设备,则会自动从 res/mipmap-mdpi 目录下获得图片资源。-hdpi、-ldpi、-mdpi 称为资源后缀,没有后缀的资源目录称为默认资源目录。Android 在搜索资源时,会首先获得最符合条件的资源,若没有符合条件的资源,则会从默认资源目录中获得资源。

如果读者愿意,完全可以删除 mipmap-ldpi、mipmap-mdpi 和 mipmap-hdpi 等目录,而在 res 目录下新建一个名为 mipmap 的图片资源目录,程序也可以正常运行。

对于图片资源文件,可以是任何目前 Android 支持的图片文件,包括.jpg 文件、.png 文件、.bmp 文件等。我们可以将需要使用的图片文件直接复制到 res/mipmap 目录下,如图 3-7 所示,我们在 mipmap 目录下放置了一个新的名称为 png2030.png 的图片。

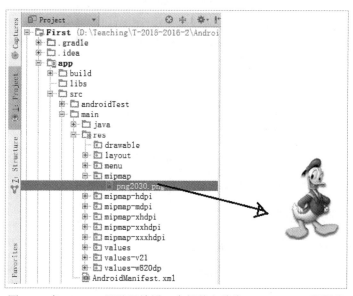

图 3-7 在 mipmap 目录下放置一个新的名称为 png2030.png 的图片

但是,我们如何使用已经添加到 Android 工程中的图片资源呢?最直接的方式就是将这个图片资源在屏幕上显示出来:我们将它显示在例子程序的 TextView 的下方。为此,我们需要对布局资源 res/layout/activity_main.xml 文件稍做修改,代码如下:

```
<LinearLayout xmlns:android="http://schemas.android.com/apk/res/android"
    android:layout_width="match_parent"
    android:layout_height="match_parent">
```

```xml
<TextView
    android:id="@+id/tv01"
    android:layout_width="wrap_content"
    android:layout_height="wrap_content"
    android:text="@string/hello_world" />

<TextView
    android:id="@+id/tv02"
    android:layout_width="wrap_content"
    android:layout_height="wrap_content"
    android:text="@string/my_message" />

<ImageView                              //新增加了一个用于显示图片的组件
    android:id="@+id/img"
    android:contentDescription=""
    android:src="@drawable/png2030"
    android:scaleType="centerInside"
    android:layout_width="128dip"
    android:layout_height="128dip" />
</LinearLayout>
```

运行修改后的程序，结果如图 3-8 所示。

图 3-8　显示图片的程序运行结果

我们将在后续章节中详细介绍常用组件的使用方法。

3.3.5　Android 的其他资源

除了上面介绍的资源，Android 中还有多种其他类型的资源，包括 color、animation、array、color-drawable 和 raw 等，后续内容将对这些资源的定义和使用方法进行详细介绍。

3.3.6　引用资源

不管 Android 资源的类型如何，所有的 Android 资源都是通过与之关联的标识符来进行引用的。例如，为了在 Java 代码中引用 message_1 字符串资源，可以使用如下语句：

```
R.string.message_1
```

来引用；而为了引用 png2030.png 图片资源，可以使用如下语句：

```
R.drawable.png2030
```

来引用。类似地，为了在资源配置 XML 文件中引用 Android 资源，例如，为了引用 message_1 字符串资源，可以使用如下语句：

`@string/message_1`

来引用；而为了引用 png2030.png 图片资源，可以使用如下语句：

`@drawable/png2030`

来引用。

一般地，在 Java 程序代码中，可以使用如下格式来引用 Android 程序资源：

`[package.]R.type.name`

类似地，在 XML 资源配置文件中，可以使用如下格式来引用 Android 程序资源：

`@[package:]type/name`

其中的 type 对应 R 文件中的资源类型，包括如下内容。

- drawable：图片资源。
- id：在布局中为组件赋予的 ID。
- layout：布局资源。
- string：字符串资源。
- menu：菜单资源。
- string-array：字符串数组资源。

name 就是为各个资源起的名字，在 R 文件中，它用一个整型常量表示。这里需要再次强调的是，资源的名称可能是一个具体的文件名，例如，布局资源、图片资源，也可能是某个资源文件中为具体资源赋予的名称，如字符串资源。

在引用 Android 资源时，若没有指明 package，则默认引用程序中自定义的资源，也就是通过 R 文件中的常量所引用的资源。在 Android 平台中，已经定义了许多的资源供用户使用，这可以从 Android 的开发文档中看到，该文档在安装 Android SDK 时，会被下载到 Android 的 SDK 目录的 docs 子目录下，打开这个目录的 index.html，即可打开帮助文档，展开 Developers，并点击 Reference，即可看到如图 3-9 所示的内容。

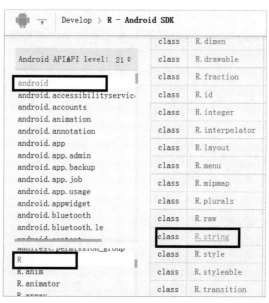

图 3-9　Android 平台自带的资源

从图 3-9 中可以看到，在 android 包下，有一个 R 类，在 R 类中有许多嵌套的内部类，打开任意一个嵌套类，如 android.R.string，即可看到如图 3-10 所示的内容。

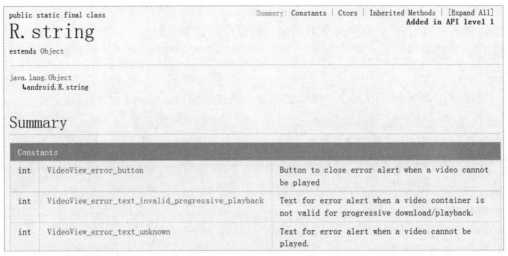

图 3-10　android.R.string 类

其中有许多字符串资源，例如：VideoView_error_button 代表一个字符串常量。在 Java 代码中，可以使用如下语句：

```
android.R.string.VideoView_error_button
```

来引用这个资源；类似地，在 XML 资源文件中，可以使用如下语句：

```
@android:string/VideoView_error_button
```

来引用这个资源。

3.4　同步练习

1．编写一个 Android 应用程序，并显示一张自己认为漂亮的图片，然后以文本的形式介绍图片的内容。

2．从 Android 的帮助文档中找出 Android 平台定义的所有图片资源，并在程序中显示其中任意一张图片。

第 4 章

深入分析 Activity

4.1 Activity 的生命周期

如前所述，Activity 是 Android 的重要组成部分，它代表的是 Android 应用程序的界面。我们在编写 C 语言或 Java SE 桌面应用程序时，都会有一个 main()函数，这个 main()函数就是程序运行的入口。我们之前编写的 Hello World 程序中的 MainActivity.java 的 Activity 代码，如下所示：

```
package com.ttt.first;

import android.os.Bundle;
import android.support.v7.app.AppCompatActivity;
import android.view.View;

public class MainActivity extends AppCompatActivity {

    @Override
    protected void onCreate(Bundle savedInstanceState) {
        super.onCreate(savedInstanceState);
        setContentView(R.layout.activity_main);
    }

}
```

查看代码我们发现其中并没有 main()函数。问题是，其没有 main()函数，Android 程序又是如何被运行的呢？这就需要我们理解程序生命周期的概念。

对生命周期这个词，其实我们都不陌生，任何有生命的东西都有生命周期，举例来说，一棵树是有生命周期的：一颗种子→发芽→长成小树→长成大树→死亡。一条鱼也是有生命周期的：鱼卵→小鱼→成年鱼→死亡。同理，一个应用程序也是有生命周期的：启动程序→完成业务功能→程序运行结束。同时还要注意的是，每个事物的生命周期都被某种其他事务管理着：小树的生命周期被大自然管理着，小鱼的生命周期也被大自然管理着，应用程序的生命周期被计算机系统管理着。回到 Android 应用程序的 Activity 界面，它也是有生命周期的：Android 平台为显示 Activity 界面做准备→显示 Activity 界面→用户与界面交互→退出 Activity 界面。其实，Android 的一个 Activity 的生命周期要比上面所描述的复杂和细致得多，并且，Android 应用程序的生命周期是由 Android 平台管理的。Android 的 Activity 的生命周期的完整过程如图 4-1 所示。

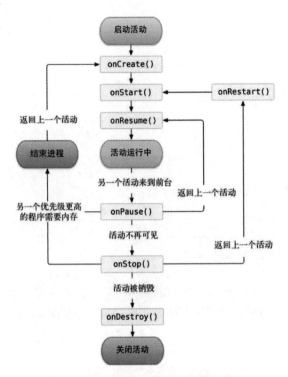

图 4-1 Android 的 Activity 的生命周期

图 4-1 中所显示的 onXXX()称为生命周期方法或生命周期状态：Android 平台会自动调用 Activity 提供的这些生命周期方法使之在适当的时候处于适当的状态。这句话的含义有些难理解，没有关系，下面我们会对它进行详细的解释。

首先，从图 4-1 可以看出，一个 Activity 从启动运行到最后的停止要经过多个状态，期间，Android 系统将根据 Activity 状态的变化，调用 Activity 的特定的生命周期方法。

当 Android 系统要启动一个应用程序的 Activity 时，例如，启动某个 Android 应用程序的入口 Activity 时，它将首先调用该 Activity 的 onCreate()方法，然后调用该 Activity 的 onStart()方法，接着调用 Activity 的 onResume()方法，在调用了 onResume()方法之后，该 Activity 已经将所表示的界面显示在屏幕上，此时，用户可以与此 Activity 进行交互，例如对于打电话的应用，则可以输入要拨打的电话号码。

当一个应用程序的 Activity 显示在屏幕上时，随时都有可能被另外一个 Activity 覆盖（也就是另外一个界面），例如，当进行语音通话时，可能突然会收到一条短信，这时，可以在继续通话的同时，打开短信应用程序来阅读收到的短信，此时，语音通话应用的 Activity 被短信应用的 Activity 覆盖。在这种情况下，Android 系统会自动调用语音通话程序 Activity 的 onPause()方法，然后再调用短信程序的入口 Activity 的 onCreate()方法、onStart()方法、onResume()方法，此时，就可以查看短信了；如果语音通话的 Activity 界面被完全覆盖，则还会调用语音通话 Activity 的 onStop()方法。当阅读完短信，并关闭短信程序后，Android 会根据语音通话 Activity 被覆盖的情况（部分或完全被短信界面覆盖），自动调用语音通话程序 Activity 的 onResume()方法，或先调用 onRestart()方法，再调用 onStart()方法，并继续调用 onResume()方法，将语音通话的 Activity 在屏幕上显示出来，这时，语音通话 Activity 的界面又显示在屏幕上了。

当用户关闭一个 Activity，即调用 Activity 的 finish()方法时，Android 会自动顺序调用该 Activity 的 onPause()→onStop()→onDestroy()来结束 Activity。例如，当打完电话要退出打电话程

序界面时，Android 将自动顺序调用打电话程序界面 Activity 的 onPause()→onStop()→onDestroy() 来结束打电话程序。

如果上面的文字介绍还是有点抽象，下面我们举一个例子来看看 Activity 的生命周期现象。

4.2 Activity 生命周期示例

我们修改之前的 MainActivity.java 程序来观察一下 Activity 生命周期的变迁。将 MainActivity.java 代码修改为如下代码：

```java
package com.ttt.first;

import android.os.Bundle;
import android.support.v7.app.AppCompatActivity;
import android.view.View;

public class MainActivity extends AppCompatActivity {

    @Override
    protected void onCreate(Bundle savedInstanceState) {
        super.onCreate(savedInstanceState);
        setContentView(R.layout.activity_main);

        System.out.println("onCreate() called");
    }

    protected void onStart() {
        super.onStart();
        System.out.println("onStart() called");
    }

    @Override
    protected void onResume() {
        super.onResume();
        System.out.println("onResume() called");
    }

    @Override
    protected void onPause() {
        super.onPause();
        System.out.println("onPause() called");
    }

    @Override
    protected void onStop() {
        super.onStop();
        System.out.println("onStop() called");
    }

    @Override
    protected void onRestart() {
        super.onRestart();
```

```
        System.out.println("onRestart() called");
    }
}
```

在修改后的代码中，我们在每个生命周期方法中加入了一些显示语句来输出一些信息到屏幕上，以便观察 Activity 的哪些生命周期方法被调用了。启动程序并运行，即可显示如图 4-2 所示的界面。

图 4-2　修改后的程序运行结果界面

为了观察程序显示的信息，打开 Android Studio 的 Android Monitor 视图：单击 Android Studio 右下方的标签，如图 4-3 所示。

图 4-3　打开 Android Studio 的 Android Monitor 视图

图 4-4 显示了 Android Monitor 的信息。

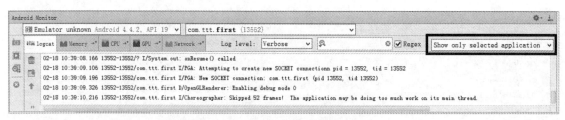

图 4-4　Android Monitor 的信息显示

此时在 Android Monitor 中显示的信息多而杂乱。为了只显示我们需要的信息，单击图 4-4 中所框住部分的下拉按钮，选择"Edit Filter Configuration"选项，即可显示如图 4-5 所示的界面。

第 4 章 深入分析 Activity

图 4-5　Edit Filter Configuration 选项配置界面

在图 4-5 中填写相关信息后，单击"OK"按钮，此时，在 Android Monitor 中只显示了我们需要的信息，如图 4-6 所示。

图 4-6　Android Monitor 中过滤后的信息

从图 4-6 中可以清楚地看到，正如我们在前面介绍的那样，MainActivity 的 onCreate()方法、onStart()方法和 onResume()方法被顺次调用。如果退出 Hello World 程序的运行，则在 Android Monitor 视图中，将显示如图 4-7 所示的信息。

图 4-7　退出 Hello World 程序时 Android Monitor 显示的信息

正如我们介绍的那样，在退出 Hello World 的 MainActivity 界面时，MainActivity 的 onPause()方法、onStop()方法和 onDestroy()方法被顺次调用。

在 Android Monitor 视图中单击鼠标右键，在弹出的快捷菜单中选择"Clear All"命令来清除 Android Monitor 中的所有信息，然后再次启动 Hello World 程序并运行，等到程序正常运行后，点击手机上的 Home 按钮，这时，Hello World 程序仍然在运行，只是 MainActivity 的界面被桌面程序的界面覆盖，这时，Android Monitor 显示如图 4-8 所示的信息。

35

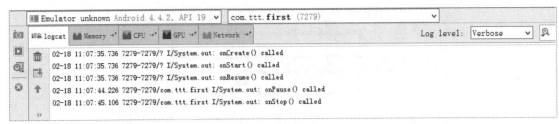

图 4-8　Hello World 的 MainActivity 界面被覆盖后 Android Monitor 显示的信息

由于 Hello World 的 MainActivity 界面完全被桌面程序覆盖，因此，MainActivity 的 onPause() 方法、onStop() 方法被顺次调用。

现在我们再次运行 Hello World 程序，也就是使 Hello World 的 MainActivity 界面再次显示出来，LogCat 将显示如图 4-9 所示的信息。

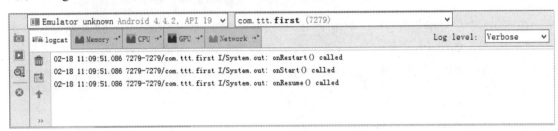

图 4-9　Hello World 的 MainActivity 界面再次显示 Android Monitor 中显示的信息

此时，MainActivity 的 onRestart()方法、onStart()方法和 onResume()方法又再次被顺次调用。

通过这个例子，我们可以清晰地看到 Activity 的生命周期方法是如何被 Android 调用的，理解这个过程是编写完美的 Activity 程序的基础。

在我们观察 Activity 生命周期过程的例子中，为了输出一些信息以便观察，我们使用了 Java 自带的 System.out.println()方法。但是，在 Android 程序中，一般建议使用 Android 的 Log 类来显示信息，以便更好地在 LogCat 中显示。下面介绍如何在程序中使用 Log 类来显示程序调试信息。

4.3　使用 Log 类输出程序调试信息

在前面我们采用 System.out 的方式输出程序调试信息，在 Android 程序中一般不建议这么做，一般建议使用 android.util.Log 类。为此，我们修改 MainActivity.java 程序，采用 Android 中建议的方式来输出调试或日志信息，修改后的代码如下：

```
package com.ttt.first;

import android.os.Bundle;
import android.support.v7.app.AppCompatActivity;
import android.util.Log;
import android.view.View;

public class MainActivity extends AppCompatActivity {
    private static String TAG = "我的调试信息";

    @Override
    protected void onCreate(Bundle savedInstanceState) {
```

```java
        super.onCreate(savedInstanceState);
        setContentView(R.layout.activity_main);
        Log.i(TAG, "onCreate() called");
    }

    protected void onStart() {
        super.onStart();
        Log.i(TAG, "onStart() called");
    }

    @Override
    protected void onResume() {
        super.onResume();
        Log.i(TAG, "onResume() called");
    }

    @Override
    protected void onPause() {
        super.onPause();
        Log.i(TAG, "onPause() called");
    }

    @Override
    protected void onStop() {
        super.onStop();
        Log.i(TAG, "onStop() called");
    }

    @Override
    protected void onRestart() {
        super.onRestart();
        Log.i(TAG, "onRestart() called");
    }

    @Override
    protected void onDestroy() {
        super.onDestroy();
        Log.i(TAG, "onDestroy() called");
    }
}
```

Android 建议为每个 Activity 分配一个 TAG，内容如下：

```
private static String TAG = "我的调试信息";
```

然后，使用 Log 类提供的静态方法显示调试信息，例如，为了在 LogCat 中显示 onCreate 信息，我们使用如下语句：

```
Log.i(TAG, "onCreate");
```

来显示 onCreate 信息。运行新修改的程序，并在 LogCat 中创建一个新的名称为"我的调试信息"的信息过滤器，在 Android Monitor 视图中将显示如图 4-10 所示的信息。

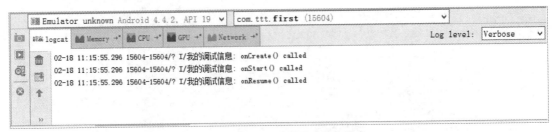

图 4-10　使用 Log 类显示程序信息

Log 类是经常用于输出调试信息的一个 Android 常用类，它提供一系列的静态方法，常用的方法及其功能如表 4-1 所示。

表 4-1　Log 类的常用方法及其功能

方法名声明	功 能
static int v(String tag,String msg)	用 VERBOSE 方式来进行消息的打印
static int v(String tag,String msg, Throwable tr)	用 VERBOSE 方式来进行消息的打印并输出异常信息
static int d(String tag,String msg)	用 DEBUG 方式来进行消息的打印
static int d(String tag,String msg, Throwable tr)	用 DEBUG 方式来进行消息的打印并输出异常信息
static int i(String tag,String msg)	用 INFO 方式来进行消息的打印
static int i(String tag, String msg, Throwable tr)	用 INFO 方式来进行消息的打印并输出异常信息
static int w(String tag, Throwable tr)	用 WARN 方式来进行异常对象的打印
static int w(String tag, String msg)	用 WARN 方式来进行消息的打印
static int w(String tag,String msg, Throwable tr)	用 WARN 方式来进行消息的打印并输出异常信息
static int e(String tag, String msg)	用 ERROR 方式来进行消息的打印
static int e(String tag,String msg, Throwable tr)	用 ERROR 方式来进行消息的打印并输出异常信息
static String getStackTraceString(Throwable tr)	取得异常对象的堆栈信息
static boolean isLoggable(String tag, int level)	指定标签是否可以在指定日志等级中进行输出
static int println(int priority,String tag, String msg)	将指定的日志消息按指定的日志等级进行输出
static int wtf(String tag,Throwable tr)	打印非常严重并永远不会发生的错误异常
static int wtf(String tag,String msg)	打印非常严重并永远不会发生的错误信息
static int wtf(String tag, String msg, Throwable tr)	打印非常严重并永远不会发生的错误信息和异常

4.4　Android 常见 Activity

Activity 及其子类是 Android SDK 中的关键类，这些类及其子类的继承关系如图 4-11 所示。

在后续章节将介绍这些类的使用方法。

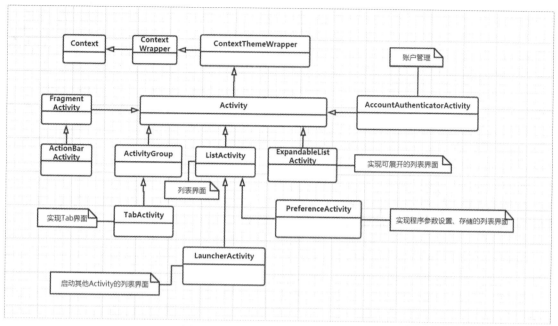

图 4-11　Activity 及其子类的继承关系

4.5　同步练习

编写一个简单的程序，将 Activity 生命周期方法的调用过程写入一个文件中，以便查看 Activity 的生命周期过程。

第 5 章

Android 常用 UI 组件

5.1 使用基于 XML 的布局

Android 建议采用将程序界面与程序业务逻辑分离的方式来进行程序设计，其实，这不是 Android 的首创。在进行网页程序设计时，我们也采用将要显示的内容与页面的布局分离的方式：将要显示的信息保存在 HTML 中，而在 CSS 中定义信息显示的格式。我们通过一个例子来展示 Android 基于 XML 的界面布局。

下面我们将新建一个 Android 程序，这个程序首先显示一个巨大的按钮，当用户点击这个按钮时，在按钮上显示当前的日期和时间。为此，在 Android Studio 中，新建一个名称为"Ex05_XML_Layout"的 Android 工程，其工程结构如图 5-1 所示。

图 5-1　Ex05_XML_Layout 工程结构

打开 activity_main.xml 文件，并修改其中的内容为如下代码：

```
<?xml version="1.0" encoding="utf-8"?>
<LinearLayout xmlns:android="http://schemas.android.com/apk/res/android"
```

```
    android:layout_width="match_parent"
    android:layout_height="match_parent">

    <Button
        android:id="@+id/id_button_01"
        android:layout_width="match_parent"
        android:layout_height="match_parent"
        android:text="@string/text_button_01" />

</LinearLayout>
```

这个布局表示在 LinearLayout 布局容器中包含一个按钮，这个按钮的 ID 为 id_button_01，该按钮将占据布局容器的整个可用显示区域，并在按钮上显示由@string/ text_button_01 引用的字符串。由于我们引用了标识符为 text_button_01 的字符串资源，因此，需要修改 res/values/strings.xml 的内容，在其中添加一个 name 为 text_button_01 的字符串资源定义，修改后的内容如下：

```
<resources>
    <string name="app_name">Ex05_XML_Layout</string>
    <string name="text_button_01">点击</string>
</resources>
```

在这个文件中，我们定义了 text_button_01 的字符串常量。运行该程序，将显示如图 5-2 所示的结果。

图 5-2　Ex05_XML_Layout 首次运行结果

在上述程序中，按钮可以被点击，在点击时按钮的颜色会发生变化，除此之外，没有其他反应。而程序的设计思路为：当点击按钮时，在按钮上显示当前的日期和时间。也就是说，程序需要响应用户对按钮的点击。那么如何来响应对按钮的点击呢？通过实现 View.OnClickListener 来达到这个目的。

View.OnClickListener 是 android.view.View 类的一个内部接口（见图 5-3），它定义了一个组件被点击时的响应函数：onClick()方法。这可以从 Android 的帮助文档中看到（如果想要快速地查看需要的 Android 类的信息，读者可以在 Android 帮助文档首页的搜索栏中直接输入想要查看的类的名称，即可查看相应类的信息。例如，为了快速查看 OnClickListener 接口的信息，在输入栏输入 OnClickListener，将直接显示该接口的信息）。

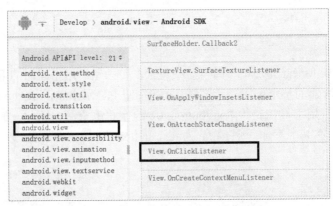

图 5-3　View.OnClickListener 接口定义

在这个接口中，只有一个方法：onClick(View v)。它表示当指定的组件被点击时的具体响应。我们对 MainActivity.java 程序代码进行修改：需要在 ID 为 id_button_01 的按钮上监听点击事件，当按钮被点击时，对这个点击事件进行适当的响应，修改后的代码如下：

```
package com.ttt.ex05_xml_layout;

import android.os.Bundle;
import android.support.v7.app.AppCompatActivity;
import android.view.View;
import android.widget.Button;

import java.util.Date;

public class MainActivity extends AppCompatActivity implements View.OnClickListener {
    Button btn;

    @Override
    protected void onCreate(Bundle savedInstanceState) {
        super.onCreate(savedInstanceState);
        setContentView(R.layout.activity_main);
        btn = (Button)this.findViewById(R.id.id_button_01);//获取界面上Button（按钮）的引用
        btn.setOnClickListener(this);                      //设置按钮监听点击事件
    }

    public void onClick(View v) {
        Date d = new Date();
        btn.setText(d.toString());
    }
}
```

MainActivity 实现了 View.OnClickListener 接口，同时，在 onCreate()方法中，程序获取界面上的 Button（按钮）的引用，并设置该按钮监听点击事件，内容如下：

```
btn = (Button)this.findViewById(R.id.id_button_01);//获取界面上Button（按钮）的引用
btn.setOnClickListener(this);                       //设置按钮监听点击事件
```

当该按钮被点击时，将调用 MainActivity 类的 onClick()方法（因为 MainActivity 类实现了 OnClickListener 接口），在 onClick()接口方法的实现中，程序获取当前的日期和时间，并将它转换为字符串显示在按钮上，内容如下：

```
public void onClick(View v) {
    Date d = new Date();
    btn.setText(d.toString());
}
```

运行修改后的程序，并点击按钮，将显示如图 5-4 所示的结果。

图 5-4　点击按钮后显示的结果

可以不停地点击按钮，每次点击按钮，都将显示当前的日期和时间。

在这个例子中，虽然我们没有对 Android 的组件进行详细的解释，但是，通过这个例子可以看出 Android 平台是如何实现布局与程序代码分离的。下面，到了该详细介绍 Android 各个界面组件及其使用方法的时候了。通过对 Android 的基本组件的介绍，我们将逐步编写更为完善的 Android 程序。

5.2 Android 基本组件

Android 基本组件包括 Button、TextView、ImageView、EditText、CheckBox、RadioButton 等。通过对这些组件的介绍，读者可以更好地掌握这些组件及其他组件的使用方法。Android 的组件在 Android.widget 包中。

5.2.1 Button

Button 组件是常用的组件，就像我们之前所举的每次点击都将显示最新日期和时间的例子一样。读者可以打开 Android 的帮助文档来观察 Button 的 XML 可配置属性。对于 Button 的每个 XML 可配置属性，均有一个与之对应的 setXXX()方法，这说明，既可以使用 XML 配置属性来配置组件的属性，也可以在代码中使用 setXXX()函数来设置同样的属性值。下面对 Button 组件的常用属性进行介绍。

Button 组件没有自己的 XML 配置属性，它的 XML 配置属性都是从它的父类 TextView 中继承来的（将在下一节介绍 TextView 组件）。Button 组件的常用布局属性包括如下内容。

（1）android:text。

设置在 Button 上要显示的文字。可以是一个字符串常量也可以是对一个字符串资源的引用。按照 Android 的规则，最好不要直接使用字符串常量，而应该使用字符串资源引用。

（2）android:textColor。

设置显示在 Button 上的文字的颜色。可以是对一个颜色资源的引用，也可以是这些形式的颜色值："#rgb"、"#argb"、"#rrggbb"或"#aarrggbb"。建议使用颜色资源引用，至于如何定义及引用颜色资源，我们会在第 6 章中进行介绍。

（3）android:textSize。

设置显示在 Button 上的文字的大小。可以是一个常数加单位，如 15px、20sp 等，也可以是对单位度量资源的引用。建议使用单位度量引用，至于如何定义及引用度量资源，我们会在后续章节进行介绍。

（4）android:textStyle。

设置显示在 Button 上的文字的风格，可用的值包括 bold、italic、bolditalic。

（5）android:typeface。

设置显示在 Button 上的文字的字体。目前 Android 只支持如下的字体：normal、sans、serif、monospace。

以上介绍的这些属性是从 android.widget.TextView 中继承来的，除此之外，还可以使用从 android.view.View 中继承的 XML 配置属性。下面来介绍这些属性。

（1）android:id。

设置 Button 的 ID 属性以便在程序代码中可以引用该组件。

（2）android:background。

设置 Button 的背景。可以是对一个 drawable 资源的引用，也可以是形如"#rgb""#argb" "#rrggbb""#aarrggbb"的颜色值。

（3）android:clickable。

设置按钮是否可以响应点击事件。可选值包括 true、false。

(4) android:visibility。

设置按钮是否显示在屏幕上。可选值包括 true、false。

(5) android:padding、android:paddingTop、android:paddingBottom、android:paddingLeft、android:paddingRight。

设置组件的内边界,类似于 HTML/CSS 中的 padding。

(6) android:gravity。

设置按钮上显示的提示文字在按钮上显示时的对齐方式。可取值包括 top、right、left、center 等。

现在修改之前所举的每次点击都显示当前日期和时间的例子,我们希望在按钮上显示一张背景图片。因此,我们需要在布局资源 activity_main.xml 文件中,为 Button 组件添加一个 android:background 属性,修改后的 res/layout/activity_main.xml 内容如下:

```xml
<?xml version="1.0" encoding="utf-8"?>
<RelativeLayout xmlns:android="http://schemas.android.com/apk/res/android"
    android:layout_width="match_parent"
    android:layout_height="match_parent">

    <Button
        android:id="@+id/id_button_01"
        android:layout_width="match_parent"
        android:layout_height="match_parent"
        android:text="@string/text_button_01"
        android:background="@mipmap/png0030"
        />

</RelativeLayout>
```

在 Button 组件的配置属性中,我们指定了按钮的背景,也就是 android:background 配置属性所指定的值:它指定采用一个名称为 png0030.png 的图片资源作为该按钮的背景图。回顾之前的介绍我们知道,需要将名称为 png0030.png 的图片文件放置到 res/mipmap-xxx 目录下,观察 Ex05_XML_Layout 工程会发现,存在多个 res/ mipmap-xxx 目录,那么,应该将图片资源文件具体放置到哪个目录下呢?

为了回答这个问题,需要知道的事实是,Android 是一个可以支持多种手持设备的操作系统平台,它既可以支持不同的手机,也可以支持平板电脑及可穿戴设备,我们在这里只讨论手机。为了使同一款程序能够在多种不同的手机上使用,需要为不同的手机(例如,不同的屏幕分辨率、不同的屏幕大小、不同的手机上使用的不同 Android 系统版本、手机是竖屏还是横屏、不同的语言等)指定不同的资源。为了给不同的手机特性指定不同的资源,Android 通过所谓的"资源配置量词"来达到这个目的。因此,res 资源目录下的所有子资源目录(例如 mipmap、drawable、layout、menu、values 等所出现的后缀名包括-ldpi、-mdpi、-hdpi、-xhdpi、-xxhdpi 等)都是资源配置量词。Android 的资源配置量词还不止这些,我们在后面会专门介绍常用的资源配置量词。

现在以在按钮上显示背景图片为例来说明如何为不同的手机指定不同的资源,也就是我们应该将 png0030.png 这个图片资源具体放置在哪个 mipmap-xxx 目录下。正确的做法是,我们应该为具有不同分辨率的手机制作不同尺寸的图片资源,为它们赋予同一个文件名,并分别放置到相应的 mipmap-mdpi、mipmap-hdpi、mipmap-xhdpi、mipmap-xxhdpi 目录下。例如,我们可以制作尺寸为 64*64 像素的图片并放置到 mipmap-mdpi 目录下、制作尺寸为 96*96 像素的图片并放置到 mipmap-hdpi 目录下、制作尺寸为 128*128 像素的图片并放置到 mipmap-xhdpi 目录下、制作尺寸为 256*256 像素的图片并放置到 mipmap-xxhdpi 目录下，这样，当程序在不同分辨率的手机上运行时所显示的图片从视觉上看大小都差不多。

我们将制作完成的不同尺寸的 png0030.png 图片资源文件分别放置到不同的 mipmap-xxx 目录下，一切就绪，现在运行修改完成的程序，结果如图 5-5 所示。

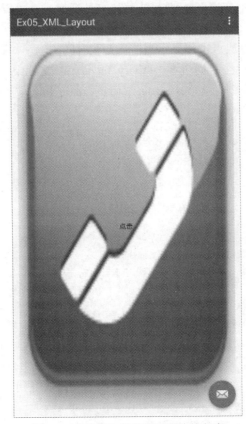

图 5-5　在按钮上显示背景图片的程序

在继续介绍 Android 的基本组件之前，我们先介绍一下 Android 的常用资源配置量词，如表 5-1 所示。

表 5-1　Android 的常用资源配置量词

类　　别	资源配置量词	含　　义
屏幕分辨率	ldpi	低密度屏幕分辨率，120dpi（120 点每英寸）
	mdpi	中密度屏幕分辨率，160dpi（160 点每英寸）
	hdpi	高密度屏幕分辨率，240dpi（240 点每英寸）
	xhdpi	超高密度屏幕分辨率，320dpi（320 点每英寸）
	xxhdpi	超超高密度屏幕分辨率，480dpi（480 点每英寸）
	xxxhdpi	超超超高密度屏幕分辨率，640dpi（640 点每英寸）
屏幕大小	small	屏幕大小至少是 320*426 dp。注意，这里的 dp 不是指屏幕的物理像素，而是指逻辑像素。dp 的英文含义是 density independent pixel，在分辨率为 160dpi 的屏幕上，1 dp = 1 px，在分辨率为 240dpi 的屏幕上，1 dp=1.5 px，以此类推

续表

类别	资源配置量词	含义
屏幕大小	normal	屏幕大小至少是 320*470 dp
	large	屏幕大小至少是 480*640 dp
	xlarge	屏幕大小至少是 720*960 dp
屏幕方位	port	手机处于竖屏状态，即手机是被竖着（Portrait）拿的
	land	手机处于横屏状态，即手机是被横着（Landscape）拿的
API 版本	v4	API 版本号为 4
	v7	API 版本号为 7
	v11	API 版本号为 11
	v14	API 版本号为 14
	等	
语言和地区	cn	中文
	en	英文
	fr	法文
	fr-rCA	法文，加拿大地区
	等	

此时再回头看前面编写的工程结构，会发现也有几个 values 的资源目录，它们分别是 values、values-v11、values-v14，这就很好理解了：在 API 版本号为 11 的 Android 手机上，使用 values-v11 资源目录下的资源；在 API 版本号大于 14 的 Android 手机上，使用 values-v14 资源目录下的资源；其他 API 版本号则默认使用 values 资源目录下的资源。

5.2.2 TextView

TextView 类似于其他 GUI 的 Label，用于显示一个字符串，显示在 TextView 中的字符串是不可编辑的。其实，Button 组件的 XML 配置都是从 TextView 和 View 中继承来的，因此，TextView 的属性与 Button 组件的 XML 配置属性是类似的。

举一个例子，修改程序代码，将之前的 Button 组件替换为 TextView 组件，修改后的 res/layout/activity_main.xml 文件内容如下：

```xml
<?xml version="1.0" encoding="utf-8"?>
<RelativeLayout xmlns:android="http://schemas.android.com/apk/res/android"
    android:layout_width="match_parent"
    android:layout_height="match_parent">

    <TextView
        android:id="@+id/textview"
        android:layout_width="match_parent"
        android:layout_height="match_parent"
        android:text="@string/hello_world"
        android:textColor="#A08800FF"
        android:textSize="15sp"
        android:textStyle="bold"
        android:typeface="monospace"
        android:background="@mipmap/png0030" />

</RelativeLayout>
```

修改后的 MainActivity.java 代码如下：

```
package com.ttt.ex05_xml_layout;

import android.os.Bundle;
import android.support.v7.app.AppCompatActivity;
import android.view.View;

public class MainActivity extends AppCompatActivity {

    @Override
    protected void onCreate(Bundle savedInstanceState) {
        super.onCreate(savedInstanceState);
        setContentView(R.layout.activity_main);

    }

}
```

修改后的 res/values/strings.xml，代码如下：

```
<resources>
    <string name="app_name">Ex05_XML_Layout</string>
    <string name="hello_world">你好，好好！</string>
</resources>
```

运行修改后的程序，结果如图 5-6 所示。

图 5-6　程序修改后的运行结果

5.2.3 ImageView

ImageView 组件用于显示一张图片。ImageView 常用的 XML 配置属性包括如下内容。

（1）android:maxHeight。

用于指定组件的最大高度。

（2）android:maxWidth。

用于指定组件的最大宽度。

（3）android:scaleType。

控制显示在 ImageView 中的图片应当如何改变大小以适应 ImageView 组件的大小。可用的值及其含义如下。

- center：图片位于视图中间，但不执行缩放。
- centerCrop：按照统一比例缩放图片（保持图片的尺寸比例）以便于图片的两维（宽度和高度）等于或大于相应的视图的维度。
- centerInside：按照统一比例缩放图片（保持图片的尺寸比例）以便于图片的两维（宽度和高度）等于或小于相应的视图的维度。
- fitCenter：按照比例缩放图片使之达到组件的大小，并使图片居中显示。
- fitEnd：按照比例缩放图片使之达到组件的大小，并使图片居末显示。
- fitStart：按照比例缩放图片使之达到组件的大小，并使图片居前显示。
- fitXY：不按比例缩放图片使之正好达到组件的维数。
- matrix：当绘制时使用图片矩阵变换进行缩放。

（4）android:src。

指定显示在组件中的图片。必须是对一个图片资源的引用。

（5）android:contentDescription。

设定图片的描述性文字。

5.2.4 EditText

EditText 是可编辑的文本组件，与 TextView 组件类似，只是提供了编辑功能。它的 XML 配置属性都是从 TextView 及 View 中继承来的，如下的几个 XML 配置属性将会在以后的编程中被使用到。

（1）android:autoText。

设置是否对输入的文字进行自动拼写检查。只能取值 true 或 false。

（2）android:captalize。

设置是否将输入的文字改为大写。只能取值 true 或 false。

（3）android:digits。

设置是否只能输入数字。只能取值 true 或 false。

（4）android:singleLine。

设置是否可以输入多行。只能取值 true 或 false。

（5）android:hint。

设置当输入框为空时，在输入框中显示的提示信息。

（6）android:inputType。

设置放置在 EditText 组件中的文字类型。取值包括 none、text、textCapCharacters、textCapWords、textUri、number 等。

5.2.5 CheckBox

CheckBox 组件就是我们在其他 GUI 组件中常用的"复选框",它继承了 TextView 和 View 组件的 XML 属性。该组件的常用方法如下。

(1) isChecked()。

检查该复选框是否被勾选。

(2) setChecked(Boolean checked)。

设置该复选框的选中状态。

(3) toggle()。

将复选框的状态反选。

5.2.6 RadioButton

RadioButton 组件就是我们在其他 GUI 组件中使用的单选按钮。在一般情况下,我们总是将 RadioButton 与 RadioGroup 结合使用:使得在一组 RadioButton 中只有一个可以被选中。通过 RadioGroup 来控制 RadioButton 的选中状态。常用的 RadioGroup 的方法如下。

(1) check(int rb)。

检查指定 rb 的选中状态。

(2) clearCheck()。

清除所有 RadioButton 的选中状态,因此,调用该方法后,没有 RadioButton 被选中。

(3) getCheckedRadioButtonId()。

返回被选中的 RadioButton 的 ID,若没有 RadioButton 被选中,则返回-1。

5.3 同步练习一

编写一个简单的 Android 程序,程序主界面上显示一个按钮,点击这个按钮,在这个按钮上以适合中国人阅读习惯的方式显示日期和时间。例如,显示日期和时间的格式应该是:2015 年 05 月 05 日 10:30:17。

5.4 Android 容器组件

Android 容器组件,也就是在其中放置其他组件并可以对放置在其中的组件进行布局的 Android 组件。常用的 Android 容器组件包括 LinearLayout、RelativeLayout、FrameLayout、ScrollView 及 CoordinatorLayout。其中,CoordinatorLayout 在 12.0 版本中做了较大更新。在这一节,我们对 Android 的常用容器组件进行介绍。

5.4.1 LinearLayout

LinearLayout 是线性布局组件,放置在其中的组件或者按列,或者按行的方式进行顺序布局。下面介绍 LinearLayout 的常用 XML 配置属性。

(1) android:orientation。

设置 LinearLayout 容器布局组件的方式:或者按行,或者按列。只能取值:horizontal、vertical。

(2) android:gravity。

设置布局在 LinearLayout 容器内的组件的对齐方式。取值包括 top、bottom、left、right、

center、start、end 等。

（3）其他从 View 中继承来的属性，包括 android:backgroud、android:visibility 等。

同时，对于布局在 LinearLayout 中的组件，LinearLayout 也提供如下这些 XML 配置属性，用以告知 LinearLayout 如何放置这些组件。这些配置属性在容器的 LinearLayout.LayoutParams 配置属性列表中，这些属性包括如下内容。

（1）android:layout_width 和 android:layout_height。

这两个属性是为放置在 LinearLayout 容器中的组件提供的：所有放置在 LinearLayout 中的组件都必须通过 android:layout_width 和 android:layout_height 属性来告知 LinearLayout 如何对组件进行布局。有 3 个可选的值：match_parent/fill_parent——占满父容器的所有空间；wrap_content——将只占用正确显示器内容所需的空间；一个常数值和单位，如 100px，表示该组件占用 100 个物理像素。常数值可用的单位包括如下内容。

- px（物理像素）：屏幕上的物理点。
- in（英寸）：长度单位。
- mm（毫米）：长度单位。
- pt（磅）：1/72 英寸。
- dp（与密度无关的像素）：一种基于屏幕密度的抽象单位。在 160 点每英寸的显示器上，1 dp = 1 px；在 240 点每英寸的显示器上，1 dp = 1.5 px。
- dip：与 dp 相同，多用于 Google 示例中。
- sp（与刻度无关的像素）：与 dp 类似，但是可以根据用户的字体大小首选项进行缩放。

（2）android:layout_gravity。

设置组件在容器中的布局方式。

（3）android:layout_weight。

设置组件占用容器的空余显示空间的比例。

（4）android:layout_margin、android:layout_marginTop、android:layout_marginBottom、android:layout_marginLeft、android:layout_marginRight

设置组件的外边界，类似于 HTML/CSS 中的 margin。

下面我们举例来说明如何使用 LinearLayout 布局。创建工程 Ex05LinearLayout，修改 activity_main.xml 文件，内容如下：

```xml
<?xml version="1.0" encoding="utf-8"?>
<LinearLayout xmlns:android="http://schemas.android.com/apk/res/android"
    android:layout_width="match_parent"
    android:layout_height="match_parent"
    android:orientation="vertical">

    <RadioGroup
       android:id="@+id/orientation"
       android:layout_width="wrap_content"
       android:layout_height="wrap_content"
       android:orientation="horizontal"
       android:padding="5dip" >

       <RadioButton
          android:layout_width="wrap_content"
          android:layout_height="wrap_content"
```

```xml
            android:id="@+id/horizontal"
            android:text="@string/horizontal" />

        <RadioButton
            android:layout_width="wrap_content"
            android:layout_height="wrap_content"
            android:id="@+id/vertical"
            android:text="@string/vertical" />
    </RadioGroup>

    <RadioGroup
        android:id="@+id/gravity"
        android:layout_width="match_parent"
        android:layout_height="wrap_content"
        android:orientation="vertical"
        android:padding="5dip" >

        <RadioButton
            android:layout_width="wrap_content"
            android:layout_height="wrap_content"
            android:id="@+id/left"
            android:text="@string/left" />

        <RadioButton
            android:layout_width="wrap_content"
            android:layout_height="wrap_content"
            android:id="@+id/center"
            android:text="@string/center" />

        <RadioButton
            android:layout_width="wrap_content"
            android:layout_height="wrap_content"
            android:id="@+id/right"
            android:text="@string/right" />
    </RadioGroup>

</LinearLayout>
```

在这个布局文件中，我们在 LinearLayout 布局容器中嵌套两个 RadioGroup，这两个 RadioGroup 分别用来管理各自的 RadioButton，使得在任何时候只有一个 RadioButton 被选中。注意，RadioGroup 也是一个容器组件：它是 LinearLayout 的子类，因此，可以在 RadioGroup 中放置 RadioButton 组件。为了能够在程序代码中获得对各个组件的引用，我们给每个组件都赋予了一个 ID。

然后，修改 res/values/strings.xml，内容如下：

```xml
<?xml version="1.0" encoding="utf-8"?>
<resources>

    <string name="app_name">Ex05LinearLayout</string>
    <string name="action_settings">Settings</string>

    <string name="horizontal">水平</string>
    <string name="vertical">垂直</string>
```

```xml
    <string name="left">居左</string>
    <string name="center">居中</string>
    <string name="right">居右</string>

</resources>
```

在这里，定义了在布局文件中用到的字符串资源常量。

继续修改 MainActivity.java，内容如下：

```java
package com.ttt.ex05linearlayout;

import android.os.Bundle;
import android.support.v7.app.AppCompatActivity;
import android.view.Gravity;
import android.view.View;
import android.widget.LinearLayout;
import android.widget.RadioGroup;

public class MainActivity extends AppCompatActivity implements RadioGroup.OnCheckedChangeListener {
    RadioGroup orientation;
    RadioGroup gravity;

    @Override
    protected void onCreate(Bundle savedInstanceState) {
        super.onCreate(savedInstanceState);
        setContentView(R.layout.activity_main);

        orientation = (RadioGroup) findViewById(R.id.orientation);
        orientation.setOnCheckedChangeListener(this);
        gravity = (RadioGroup) findViewById(R.id.gravity);
        gravity.setOnCheckedChangeListener(this);
    }

    public void onCheckedChanged(RadioGroup group, int checkedId) {
        switch (checkedId) {
            case R.id.horizontal:
                orientation.setOrientation(LinearLayout.HORIZONTAL);
                break;
            case R.id.vertical:
                orientation.setOrientation(LinearLayout.VERTICAL);
                break;
            case R.id.left:
                gravity.setGravity(Gravity.START);
                break;
            case R.id.center:
                gravity.setGravity(Gravity.CENTER_HORIZONTAL);
                break;
            case R.id.right:
                gravity.setGravity(Gravity.END);
                break;
        }
    }
}
```

在 MainActivity 的 onCreate()方法中，我们获得了对两个 RadioGroup 组件的引用，并分别设置了对其中的 RadioButton 选中状态发生变化时的监听，并在 OnCheckedChangeListener 接口的 onCheckedChanged()方法中对选中状态的变化进行处理，例如，若标记为"水平"的单选按钮被选中，则调用 RadioGroup 的 setOritentation()方法设置这个 RadioGroup 的方位为水平，对重心的处理与此类似，只是采用 RadioGroup 的 setGravity()方法设置重心而已。

LinearLayout 布局容器为布局于其中的组件提供了 android:layout_weight 布局属性，这个布局属性很有意思：它可以使各个组件按指定的比例来共享 LinearLayout 容器的显示空间。下面我们举例来说明如何使用 android:layout_weight 布局属性。

首先在 Eclipse 中新建一个名称为 Ex05LinearLayoutWeight 的 Android 工程，修改 res/layout/content_main.xml 为如下内容：

```xml
<?xml version="1.0" encoding="utf-8"?>
<LinearLayout xmlns:android="http://schemas.android.com/apk/res/android"
    android:layout_width="match_parent"
    android:layout_height="match_parent"
    android:orientation="vertical">

    <Button
        android:layout_width="match_parent"
        android:layout_height="0dp"
        android:layout_weight="5"
        android:text="@string/fifty" />

    <Button
        android:layout_width="match_parent"
        android:layout_height="0dp"
        android:layout_weight="3"
        android:text="@string/thirty" />

    <Button
        android:layout_width="match_parent"
        android:layout_height="0dp"
        android:layout_weight="2"
        android:text="@string/twenty" />

</LinearLayout>
```

在这个 LinearLayout 布局中包含 3 个 Button（按钮）。在采用垂直布局的情况下，使用 android:layout_weight 时，需要设置 android:layout_height 的值为 0dp；类似地，在采用水平布局的情况下，使用 android:layout_weight 时，需要设置 android:layout_width 的值为 0dp。然后，在相应组件的 android:layout_weight 属性中指定这个组件要占用的显示空间的比例，例如，第 1 个按钮所占用的显示空间为在 10 等份中占 5 等份，第 2 个按钮占 3 等份，第 3 个按钮占 2 等份。

当然，我们还需要修改 res/values/strings.xml 文件，在这个文件中，对布局中引用的 3 个字符串资源常量进行定义，内容如下：

```xml
<?xml version="1.0" encoding="utf-8"?>
<resources>

    <string name="app_name">Ex05LinearLayoutWeight</string>
    <string name="fifty">50%</string>
```

```
    <string name="thirty">30%</string>
    <string name="twenty">20%</string>
</resources>
```

运行该程序,将显示如图 5-7 所示的界面。

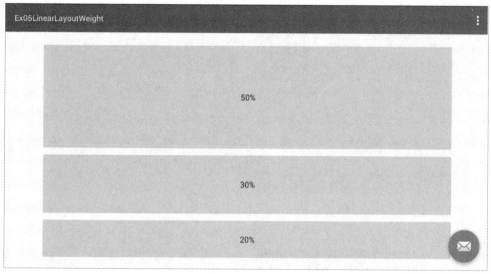

图 5-7　android:layout_weight 属性的使用

5.4.2　RelativeLayout

RelativeLayout,顾名思义,就是相对布局:一个组件相对于另一个组件的位置来构造布局。例如,将组件 A 布局在组件 B 的右下方。RelativeLayout 为布局在其中的组件提供了非常多的布局属性,表 5-2 列出了常用的 XML 布局属性及其含义。

表 5-2　RelativeLayout 常用的 XML 布局属性及其含义

XML 布局属性	含　　义
android:layout_above	定位组件到指定的组件的上方
android:layout_alignParentBottom	如果为 true 值,定位组件到父容器的最下方
android:layout_alignParentLeft	如果为 true 值,定位组件到父容器的最左边
android:layout_alignParentRight	如果为 true 值,定位组件到父容器的最右边
android:layout_alignParentTop	如果为 true 值,定位组件到父容器的最上方
android:layout_centerInParent	如果为 true 值,定位组件到父容器的中央
android:layout_below	定位组件到指定的组件的下方

现在举一个例子来说明 RelativeLayout 布局容器的使用方法。在 Android Studio 中新建一个名称为 Ex05RelativeLayout 的工程,然后修改 res/layout/content_main.xml 布局文件为如下内容:

```
<?xml version="1.0" encoding="utf-8"?>
<RelativeLayout xmlns:android="http://schemas.android.com/apk/res/android"
    android:layout_width="match_parent"
    android:layout_height="match_parent">

    <TextView
        android:id="@+id/userNameLbl"
```

```xml
        android:layout_width="match_parent"
        android:layout_height="wrap_content"
        android:layout_alignParentTop="true"
        android:text="@string/username" />

    <EditText
        android:id="@+id/userNameText"
        android:layout_width="match_parent"
        android:layout_height="wrap_content"
        android:hint=""
        android:layout_below="@id/userNameLbl" />

    <TextView
        android:id="@+id/pwdLbl"
        android:layout_width="match_parent"
        android:layout_height="wrap_content"
        android:layout_below="@id/userNameText"
        android:text="@string/password" />

    <EditText
        android:id="@+id/pwdText"
        android:layout_width="match_parent"
        android:layout_height="wrap_content"
        android:hint=""
        android:layout_below="@id/pwdLbl" />

    <TextView
        android:id="@+id/pwdCriteria"
        android:layout_width="match_parent"
        android:layout_height="wrap_content"
        android:layout_below="@id/pwdText"
        android:text="@string/criteria" />

    <TextView
        android:id="@+id/disclaimerLbl"
        android:layout_width="match_parent"
        android:layout_height="wrap_content"
        android:layout_alignParentBottom="true"
        android:text="@string/risk" />

</RelativeLayout>
```

然后再修改 res/values/strings.xml 文件，内容如下：

```xml
<?xml version="1.0" encoding="utf-8"?>
<resources>

    <string name="app_name">Ex05RelativeLayout</string>

    <string name="username">用户名</string>
    <string name="password">密码</string>
    <string name="criteria">密码规范</string>
    <string name="risk">风险</string>

</resources>
```

运行该程序，结果如图 5-8 所示。

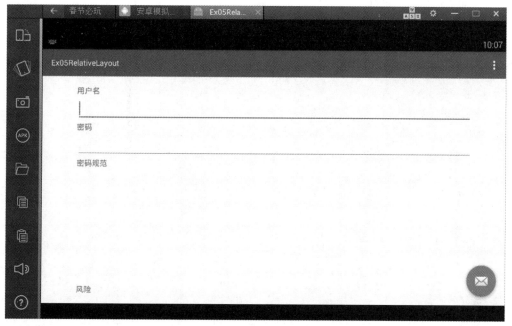

图 5-8　RelativeLayout 布局运行结果

5.4.3　FrameLayout

FrameLayout 以层叠的方式布局组件：每次只能显示其中的一个。与扑克牌类似，当叠加在一起时只能看到最上面的那张。FrameLayout 为布局在其中的组件提供了一个 XML 配置属性：android:layout_gravity。通过这个属性，布局在 FrameLayout 中的组件可以指定自己在容器中的重心位置，例如，靠左、靠右等。

举一个例子来说明 FrameLayout 的应用，在 Android Studio 中新建一个名为 Ex05FrameLayout 的 Android 工程，在程序中显示 4 张图片，需要先在 drawable 目录下放置 4 张图片资源。然后修改 res/layout/activity_main.xml 布局文件，内容如下：

```
<?xml version="1.0" encoding="utf-8"?>
<FrameLayout xmlns:android="http://schemas.android.com/apk/res/android"
    android:id="@+id/frmLayout"
    android:layout_width="match_parent"
    android:layout_height="match_parent">

    <ImageView
        android:id="@+id/id_iv01"
        android:layout_width="match_parent"
        android:layout_height="match_parent"
        android:scaleType="fitCenter"
        android:contentDescription="@string/text_empty"
        android:src="@drawable/png0066"
        android:visibility="visible" />

    <ImageView
        android:id="@+id/id_iv02"
        android:layout_width="match_parent"
```

```xml
        android:layout_height="match_parent"
        android:scaleType="fitCenter"
        android:contentDescription="@string/text_empty"
        android:src="@drawable/png0067"
        android:visibility="gone" />

    <ImageView
        android:id="@+id/id_iv03"
        android:layout_width="match_parent"
        android:layout_height="match_parent"
        android:scaleType="fitCenter"
        android:contentDescription="@string/text_empty"
        android:src="@drawable/png0068"
        android:visibility="gone" />

    <ImageView
        android:id="@+id/id_iv04"
        android:layout_width="match_parent"
        android:layout_height="match_parent"
        android:scaleType="fitCenter"
        android:contentDescription="@string/text_empty"
        android:src="@drawable/png0069"
        android:visibility="gone" />

</FrameLayout>
```

在这个布局文件中，我们在 FrameLayout 中放置 4 个 ImageView 组件，并且只显示第一个 ImageView 组件，而将其他的 ImageView 组件的 android:visibility 属性值设为"gone"，注意，我们也可以将这个值设为"invisible"。gone 与 invisible 的相同点是：它们都让组件不再显示出来；而区别在于，gone 不仅不显示组件，而且使组件也不占据显示空间。

修改 res/values/strings.xml 文件，内容如下：

```xml
<?xml version="1.0" encoding="utf-8"?>
<resources>

    <string name="app_name">Ex05FrameLayout</string>

    <string name="text_empty"></string>

</resources>
```

其中定义了在布局文件中引用的字符串常量。

修改 MainActivity.java 文件，内容如下：

```java
package com.ttt.ex05framelayout;

import android.os.Bundle;
import android.support.v7.app.AppCompatActivity;
import android.view.View;
import android.widget.ImageView;

public class MainActivity extends AppCompatActivity implements View.OnClickListener {
    private ImageView iv01,iv02,iv03,iv04;
```

```java
    @Override
    protected void onCreate(Bundle savedInstanceState) {
        super.onCreate(savedInstanceState);
        setContentView(R.layout.activity_main);
        Toolbar toolbar = (Toolbar) findViewById(R.id.toolbar);
        setSupportActionBar(toolbar);

        FloatingActionButton fab = (FloatingActionButton) findViewById(R.id.fab);
        fab.setOnClickListener(new View.OnClickListener() {
            @Override
            public void onClick(View view) {
                Snackbar.make(view, "Replace with your own action",
                        Snackbar.LENGTH_LONG)
                        .setAction("Action", null).show();
            }
        });

        iv01 = (ImageView)this.findViewById(R.id.id_iv01);
        iv01.setOnClickListener(this);
        iv02 = (ImageView)this.findViewById(R.id.id_iv02);
        iv02.setOnClickListener(this);
        iv03 = (ImageView)this.findViewById(R.id.id_iv03);
        iv03.setOnClickListener(this);
        iv04 = (ImageView)this.findViewById(R.id.id_iv04);
        iv04.setOnClickListener(this);
    }

    public void onClick(View v) {
        int id = v.getId();
        switch(id) {
            case R.id.id_iv01:
                iv01.setVisibility(View.GONE);
                iv02.setVisibility(View.VISIBLE);
                break;

            case R.id.id_iv02:
                iv02.setVisibility(View.GONE);
                iv03.setVisibility(View.VISIBLE);
                break;

            case R.id.id_iv03:
                iv03.setVisibility(View.GONE);
                iv04.setVisibility(View.VISIBLE);
                break;

            case R.id.id_iv04:
                iv04.setVisibility(View.GONE);
                iv01.setVisibility(View.VISIBLE);
                break;
        }
    }
}
```

在 MainActivity 代码的 onCreate()方法中，首先获取各个 ImageView 组件的引用，并设置它们的监听点击事件。然后，在 OnClickListener 接口的 onClick()方法中，判断是哪一个 ImageView 被点击，并根据当前被点击的 ImageView 的 ID 来显示下一个 ImageView。运行这个程序，将从第 2 张开始显示，并将一直在 4 张图片之间循环显示。

5.4.4 ScrollView

ScrollView 也是一个容器，它是 FrameLayout 的子类，它的主要作用是可以将超出物理屏幕的内容显示出来，ScrollView 提供垂直滚动，进而可将超出物理屏幕的内容显示出来。

在一般情况下，可以将一个采用垂直方式布局组件的 LinearLayout 作为 ScrollLayout 容器的子组件，同时，在 LinearLayout 容器中可以显示超出屏幕物理高度的内容。

下面举例说明 ScrollView 的使用方法。在 Android Studio 中新建一个名称为 Ex05ScrollView 的 Android 工程。

修改 res/layout/content_main.xml 布局文件，虽然代码看起来有点长，但结构很简单，就是在 ScrollView 中嵌套一个 LinearLayout，再在这个 LinearLayout 中嵌套多个 LinearLayout，在每个二级嵌套的 LinearLayout 中显示一个颜色和这个颜色的编码。activity_main.xml 文件内容如下：

```xml
<?xml version="1.0" encoding="utf-8"?>
<ScrollView xmlns:android="http://schemas.android.com/apk/res/android"
    android:layout_width="match_parent"
    android:layout_height="wrap_content">

    <LinearLayout
        android:layout_width="match_parent"
        android:layout_height="wrap_content"
        android:orientation="vertical" >

        <LinearLayout
            android:layout_width="match_parent"
            android:layout_height="wrap_content"
            android:orientation="horizontal" >

            <View
                android:layout_width="0dp"
                android:layout_height="80dp"
                android:layout_weight="4"
                android:background="#000000" />

            <TextView
                android:layout_width="0dp"
                android:layout_height="80dp"
                android:layout_weight="2"
                android:gravity="center"
                android:text="@string/text_000000" />
        </LinearLayout>

        <LinearLayout
            android:layout_width="match_parent"
            android:layout_height="wrap_content"
```

```xml
        android:orientation="horizontal" >

        <View
            android:layout_width="0dp"
            android:layout_height="80dp"
            android:layout_weight="4"
            android:background="#440000" />

        <TextView
            android:layout_width="0dp"
            android:layout_height="80dp"
            android:layout_weight="2"
            android:gravity="center"
            android:text="@string/text_440000" />
    </LinearLayout>

    <LinearLayout
        android:layout_width="match_parent"
        android:layout_height="wrap_content"
        android:orientation="horizontal" >

        <View
            android:layout_width="0dp"
            android:layout_height="80dp"
            android:layout_weight="4"
            android:background="#884400" />

        <TextView
            android:layout_width="0dp"
            android:layout_height="80dp"
            android:layout_weight="2"
            android:gravity="center"
            android:text="@string/text_884400" />
    </LinearLayout>

    <LinearLayout
        android:layout_width="match_parent"
        android:layout_height="wrap_content"
        android:orientation="horizontal" >

        <View
            android:layout_width="0dp"
            android:layout_height="80dp"
            android:layout_weight="4"
            android:background="#aa8844" />

        <TextView
            android:layout_width="0dp"
            android:layout_height="80dp"
            android:layout_weight="2"
            android:gravity="center"
            android:text="@string/text_aa8844" />
    </LinearLayout>

    <LinearLayout
```

```xml
        android:layout_width="match_parent"
        android:layout_height="wrap_content"
        android:orientation="horizontal" >

        <View
            android:layout_width="0dp"
            android:layout_height="80dp"
            android:layout_weight="4"
            android:background="#ffaa88" />

        <TextView
            android:layout_width="0dp"
            android:layout_height="80dp"
            android:layout_weight="2"
            android:gravity="center"
            android:text="@string/text_ffaa88" />
    </LinearLayout>

    <LinearLayout
        android:layout_width="match_parent"
        android:layout_height="wrap_content"
        android:orientation="horizontal" >

        <View
            android:layout_width="0dp"
            android:layout_height="80dp"
            android:layout_weight="4"
            android:background="#ffffaa" />

        <TextView
            android:layout_width="0dp"
            android:layout_height="80dp"
            android:layout_weight="2"
            android:gravity="center"
            android:text="@string/text_ffffaa" />
    </LinearLayout>

    <LinearLayout
        android:layout_width="match_parent"
        android:layout_height="wrap_content"
        android:orientation="horizontal" >

        <View
            android:layout_width="0dp"
            android:layout_height="80dp"
            android:layout_weight="4"
            android:background="#ffffff" />

        <TextView
            android:layout_width="0dp"
            android:layout_height="80dp"
            android:layout_weight="2"
            android:gravity="center"
            android:text="@string/text_ffffff" />
    </LinearLayout>
```

```
        </LinearLayout>
</ScrollView>
```

再修改 res/values/strings.xml 资源文件,在其中定义一些在 activity_main.xml 中引用到的字符串常量,strings.xml 文件的内容如下:

```xml
<?xml version="1.0" encoding="utf-8"?>
<resources>

    <string name="app_name">Ex05ScrollView</string>

    <string name="text_000000">#000000</string>
    <string name="text_440000">#440000</string>
    <string name="text_884400">#884400</string>
    <string name="text_aa8844">#aa8844</string>
    <string name="text_ffaa88">#ffaa88</string>
    <string name="text_ffffaa">#ffffaa</string>
    <string name="text_ffffff">#ffffff</string>

</resources>
```

运行该程序,结果如图 5-9 所示。由于显示的内容超出了物理屏幕的高度,自动出现了垂直滚动条。

图 5-9　ScrollView 程序运行结果界面

注意　ScrollView 只提供垂直滚动条,若要使用水平滚动功能,则 Android 提供了 HorizontalScrollView 容器,HorizontalScrollView 容器可以提供水平滚动功能,它的使用方法与 ScrollView 类似。

5.4.5　CoordinatorLayout

CoordinatorLayout 是 Android 中一个非常有用的布局容器类。请读者自行阅读 Android 的 SDK 文档进行学习,然后阅读一篇博文:http://blog.csdn.net/xyz_lmn/article/details/48055919。

5.5 同步练习二

编写一个简单的计算器程序，只需要完成基本的加、减、乘、除运算即可。要求：界面布局简洁、美观。

5.6 AdapterView

在 Android 应用开发中，AdapterView 是一类常用的且非常重要的组件。我们常见的以列表的形式显示信息的组件就是 AdapterView 的子类，称为 ListView；我们经常以网格形式浏览照片缩略图的组件也是 AdapterView 的子类，称为 GridView；以下拉列表形式显示可选项的组件也是 AdapterView 的子类，称为 Spinner；……它们都是 AdapterView 的子类。在这一节，我们对 AdapterView 进行详细介绍。

5.6.1 AdapterView 入门

在介绍 AdapterView 组件的使用之前，我们先通过一个例子来了解一下 AdapterView 的编程模式。以 ListView 为例，先看一个例子程序的界面，如图 5-10 所示。

图 5-10 系统中安装的应用的名称和图标界面

这个例子用 Android 的 ListView 组件以列表的形式显示 Android 系统中已经安装的所有程序信息。ListView，顾名思义，就是通过列表的形式向用户展示信息。在图 5-10 中，其使用 ListView 显示了 Android 系统中已经安装的所有的应用程序的图标、应用程序名称和入口 Activity 的类名，当显示的内容超出物理屏幕可用区域时，它还可以进行滚动。

在图 5-10 中，大框框住的部分是一个 ListView，小框框住的部分是列表中的一个列表项，因此，在程序中要使用 ListView 显示信息，必须做如下的工作。

（1）在界面布局中包含一个 ListView 组件。
（2）对在列表中显示的列表项进行布局。
（3）设计一个实现了 Adapter 接口的类，用于为 ListView 组件提供需要显示的数据。

基于前面对 UI 组件的介绍，将 ListView 组件纳入布局并不是一件困难的事，同时，对列表项进行布局也不是一件难事，但是要用好 ListView 或其他的 AdapterView，我们需要对 Adapter 接口有一个深入的理解。

5.6.2 Adapter

之前提到的列表组件（ListView）、网格组件（GridView）和下拉列表组件（Spinner），它们都是 AdapterView 的子类，这些组件只负责显示数据，而对于这些要显示的数据则必须通过称为 Adapter 的接口来进行管理。以使用 ListView 显示数据为例，AdapterView 和 Adapter 接口的关系如图 5-11 所示。

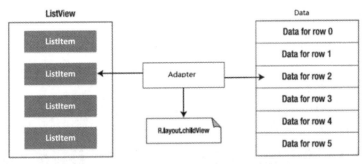

图 5-11　AdapterView 和 Adapter 接口的关系

从图 5-11 可以看出，ListView 只负责显示数据，并处理滚动、对列表项的点击等事件，在其中要显示数据则由 Adapter 接口提供。其实，Adapter 接口不仅提供数据，它还必须将要显示在 ListView 中的数据进行布局封装，形成完整的组件，并作为一个列表项提供给 ListView，然后将这个组件显示出来。因此，可以这么说，要在 AdapterView 中显示数据，大部分工作都是在 Adapter 中完成的。

那么，AdapterView 是如何与 Adapter 交互来完成信息显示功能的呢？AdapterView 显示数据的处理逻辑如下。

（1）当 AdapterView 要显示一项数据时，例如，对 ListView 而言，就是当 ListView 要显示一行数据时，它首先会调用 Adapter 的 getView() 方法，并传递一个要显示的数据的位置参数。

（2）Adapter 根据这个位置参数从 Data 中获得指定的数据，并根据 R.layout.childView 的布局样式将数据填入到样式布局中，然后，将这个构建好的 View 返回给 AdapterView。

（3）AdapterView 将这个返回的 View 作为子 View 显示在组件中。

Adapter 接口是如此的重要，所以我们需要对 Adapter 接口有一个更加深入的了解。在 Android 的开发文档中，对 Adapter 接口是这样描述的：Adapter 接口对象是 AdapterView 和要显示的数据的桥梁，它提供了对要显示的数据的访问机制，同时，Adapter 接口对象还必须为 AdapterView 生成要显示的组件并提供给 AdapterView 进行显示。

这个描述可能有点抽象，没有关系，我们会通过例子来介绍如何通过设计来实现 Adapter 接口的类。这是下一步的工作。我们现在需要对 Adapter 接口中的方法及其含义进行了解。Adapter 接口中定义的常用方法及其含义如表 5-3 所示。

表 5-3 Adapter 接口中定义的常用方法及其含义

方 法 名 称	含 义
int getCount( )	返回要显示的数据集中的数据总数
Object getItem(int position)	返回数据集中指定位置的数据对象
long getItemId(int position)	返回数据集中指定位置的数据的 ID
View getView(int position, View convertView, ViewGroup parent)	将指定位置的数据构建成一个可以显示在 AdapterView 中的组件,并返回 AdapterView 进行显示

Adapter 只是一个接口,为了方便程序设计,Android 平台 SDK 提供了几个实现 Adapter 接口的基础类,它们分别是:BaseAdapter、ArrayAdapter、SimpleAdapter、SimpleCursorAdapter 等。一般来说,我们需要对这些基础类进行扩展来实现有特殊要求的 Adapter。下面我们结合例子来介绍各个 AdapterView 及为之提供数据的 Adapter 的使用。

5.6.3 ListView

5.6.3.1 ListView 基础

ListView 以垂直的方式显示数据列表,并且,当要显示的数据多于屏幕高度时,它可以以垂直滚动的方式显示其余数据。

我们来看一个 ListView 的例子:ListView 以列表的形式显示书的名称,分别为 Java 程序设计、Android 应用开发、Oracle 数据库管理指南等。程序运行的结果如图 5-12 所示。

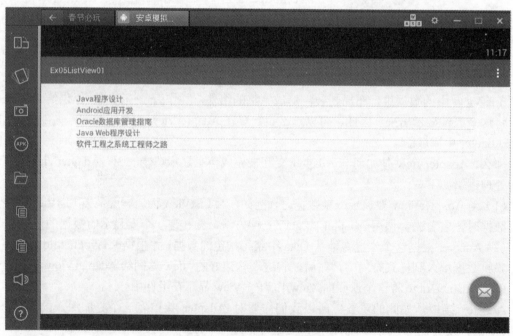

图 5-12 简单的 ListView 例子

也就是以列表的形式显示了书的名称。这个程序的源代码非常简单,我们可以通过这个简单的例子了解使用 ListView 及 Adapter 的基本方法。

首先在 Android Studio 中,新建一个名为 Ex05ListView01 的 Android 工程。还记得我们在 5.6.1 节介绍的使用 AdapterView 显示信息的 3 个步骤吗?

（1）在界面布局中包含一个 ListView 组件。
（2）对在列表中显示的列表项进行布局。
（3）设计一个实现了 Adapter 接口的类，用于为 ListView 组件提供要显示的数据。

首先完成第一步工作。我们的程序主界面是 activity_main.xml，因此，为了完成第一步工作，需要先修改主界面布局文件，修改后的布局文件 res/layout/activity_main.xml 的内容如下：

```xml
<?xml version="1.0" encoding="utf-8"?>
<RelativeLayout xmlns:android="http://schemas.android.com/apk/res/android"
    android:layout_width="match_parent"
    android:layout_height="wrap_content">

    <ListView
        android:id="@+id/id_lv"
        android:layout_width="match_parent"
        android:layout_height="match_parent"/>

</RelativeLayout>
```

在主界面的布局文件中，我们在布局管理器中包含一个 ListView 组件，并且让这个 ListView 在水平方向占满整个空间，在垂直方向按内容多少来占用显示空间。

其次来完成第二步工作：为列表项创建一个布局。为此，在工程目录的 res/layout 目录下，单击鼠标右键，在弹出的快捷菜单中，选择 New→Layout resource file，如图 5-13 所示。

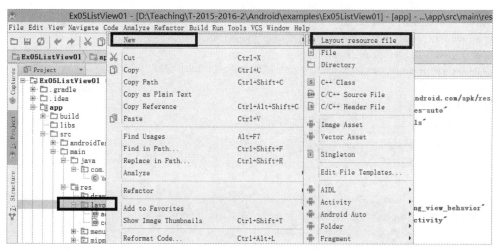

图 5-13　新建一个布局文件

在弹出窗口的 File 文本框中填写 list_item.xml，单击 Finish 按钮，并修改 list_item.xml 文件内容如下：

```xml
<?xml version="1.0" encoding="utf-8"?>
<TextView xmlns:android="http://schemas.android.com/apk/res/android"
    android:id="@+id/id_book_name"
    android:layout_width="match_parent"
    android:layout_height="match_parent"
/>
```

这个布局是列表框中列表项的布局：每个列表项都只是采用 TextView 组件来显示书的名称。

最后完成第三步工作：编写自己的 Adapter。在 Ex05ListView01 工程中，新建一个名为 MyAdapter 的 Java 类。操作顺序如图 5-14 所示。

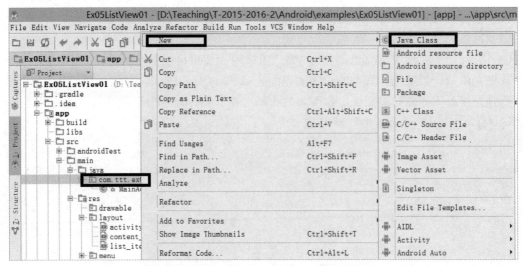

图 5-14 新建 MyAdapter 的 Java 类的操作顺序

修改 MyAdapter.java 文件，内容如下：

```java
package com.ttt.ex05listview01;

import android.annotation.SuppressLint;
import android.content.Context;
import android.view.LayoutInflater;
import android.view.View;
import android.view.ViewGroup;
import android.widget.BaseAdapter;
import android.widget.TextView;

public class MyAdapter extends BaseAdapter {
    private String[] books = {"Java 程序设计", "Android 应用开发", "Oracle 数据库管理指南",
            "Java Web 程序设计", "软件工程之系统工程师之路"};

    LayoutInflater inflater;
    int id_list_item;

    public MyAdapter(Context context, int id_list_item) {
        this.id_list_item = id_list_item;
        inflater = (LayoutInflater)
                context.getSystemService(Context.LAYOUT_INFLATER_SERVICE);
    }

    @Override
    public int getCount() {
        return books.length;
    }

    @Override
    public Object getItem(int position) {
        return books[position];
    }

    @Override
```

```
    public long getItemId(int position) {
        return position;
    }

    @SuppressLint("ViewHolder") @Override
    public View getView(int position, View convertView, ViewGroup parent) {
        TextView tv;
        tv = (TextView)inflater.inflate(id_list_item, parent, false);
        tv.setText(books[position]);

        return tv;
    }
}
```

如前所述，编写自己的 Adapter 是使用 AdapterView 的关键。为了在 ListView 中显示书的名称，我们创建了自己的 Adapter 接口的实现类：MyAdapter。为了简化设计，MyAdapter 直接继承 BaseAdapter，注意，BaseAdapter 只是 Adapter 接口的简单实现，它并没有做什么工作，根据需要，我们重载了其中的几个重要方法。其中的 getCount()和 getView()方法的实现需要在此着重解释。

我们前面已经介绍过，当 ListView 要显示信息时，它会首先调用它所对应的 Adapter 的 getCount()方法来询问在 Adapter 中共有多少条数据需要显示，以便使 ListView 做一些准备工作。在 MyAdapter 类中，返回 books 数组的条目给 ListView，当 ListView 要显示具体的数据时，系统会调用 Adapter 接口的 getView()方法以获得要显示的信息。因此，在 getView()方法中，我们先使用布局展开器将列表项布局文件 R.layout.list_item 展开成视图。同时，因为 list_item.xml 布局文件只是一个简单的 TextView，因此，我们可以直接将其强制转换成 TextView 对象。然后将对应的 books 数组中的书的名称显示在该 TextView 组件中，并将这个组装好的组件返回给 ListView，从而 ListView 可将该组件在列表中显示出来。

完成以上工作后，现在要做的就是进行简单的整合。为此，修改 MainActivity.java 代码，在其中的 onCreate()方法中获得主界面的 ListView 组件，并告诉这个 ListView 组件从 MyAdapter 这个 Adapter 中获得要显示的数据，修改后的 MainActivity.java 代码如下：

```
package com.ttt.ex05listview01;

import android.os.Bundle;
import android.support.design.widget.FloatingActionButton;
import android.support.design.widget.Snackbar;
import android.support.v7.app.AppCompatActivity;
import android.support.v7.widget.Toolbar;
import android.view.View;
import android.view.Menu;
import android.view.MenuItem;
import android.widget.ListView;

public class MainActivity extends AppCompatActivity {

    @Override
    protected void onCreate(Bundle savedInstanceState) {
        super.onCreate(savedInstanceState);
        setContentView(R.layout.activity_main);
```

```
Toolbar toolbar = (Toolbar) findViewById(R.id.toolbar);
setSupportActionBar(toolbar);

FloatingActionButton fab = (FloatingActionButton) findViewById(R.id.fab);
fab.setOnClickListener(new View.OnClickListener() {
    @Override
    public void onClick(View view) {
        Snackbar.make(view, "Replace with your own action",
                Snackbar.LENGTH_LONG)
                .setAction("Action", null).show();
    }
});

ListView lv = (ListView)this.findViewById(R.id.id_lv);    //获得ListView组件
MyAdapter mad = new MyAdapter(this, R.layout.list_item);
lv.setAdapter(mad);    //告诉ListView从mad这个Adapter中获得数据
    }
}
```

注意 onCreate()方法中的代码：

```
ListView lv = (ListView)this.findViewById(R.id.id_lv);    //获得ListView组件
MyAdapter mad = new MyAdapter(this, R.layout.list_item);
lv.setAdapter(mad);    //告诉ListView从mad这个Adapter中获得数据
```

这段代码的含义是获得 ListView 组件的引用，并告诉这个 ListView 从 mad 这个 Adapter 中获得数据。

运行完成的 Ex05ListView01 程序，即可得到如图 5-12 所示的结果。

5.6.3.2 使 ListView 显示的信息更漂亮

我们查看上一个 ListView 中显示的书的名称，可以说效果太单调了。下面要做的工作就是使显示的信息更漂亮。

修改上面的程序，并希望达到如下的运行效果：在 ListView 的每 1 行，除了显示书的名称，我们还希望显示该书的封面图片，如图 5-15 所示。

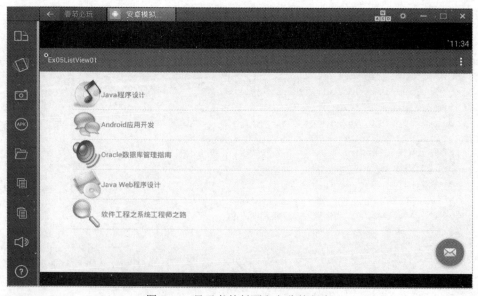

图 5-15　显示书的封面和名称的程序

为了实现这个显示效果，我们需要修改列表项的布局文件和 MyAdapter 类文件。首先修改列表项布局文件。

显然，现在的列表项不仅需要显示书的名称的 TextView，还需要显示书的封面的 ImageView，为此，修改 res/layout/list_item.xml 文件，内容如下：

```xml
<?xml version="1.0" encoding="utf-8"?>
<LinearLayout xmlns:android="http://schemas.android.com/apk/res/android"
    android:layout_width="match_parent"
    android:layout_height="match_parent"
    android:orientation="horizontal" >

    <ImageView
        android:id="@+id/id_book_photo"
        android:layout_width="60dp"
        android:layout_height="60dp"
        android:contentDescription="@string/text_empty"    //不要产生警告
    />

    <TextView
        android:id="@+id/id_book_name"
        android:layout_width="match_parent"
        android:layout_height="match_parent"
        android:gravity="center_vertical"
    />

</LinearLayout>
```

在列表项的布局中，我们增加了一个用于显示图片的 ImageView。注意，为了使 ImageView 不产生警告，我们在 ImageView 的 android:contentDescription 属性中为之赋予了一个 text_empty 的字符串引用，为此，需要在 res/values/strings.xml 文件中增加一行字符串定义，修改后的 strings.xml 文件内容如下：

```xml
<?xml version="1.0" encoding="utf-8"?>
<resources>

    <string name="app_name">Ex05ListView01</string>
    <string name="action_settings">Settings</string>

    <string name="text_empty"></string>

</resources>
```

再修改 MyAdapter.java 文件，内容如下：

```java
package com.ttt.ex05listview01;

import android.annotation.SuppressLint;
import android.content.Context;
import android.view.LayoutInflater;
import android.view.View;
import android.view.ViewGroup;
import android.widget.BaseAdapter;
import android.widget.ImageView;
import android.widget.LinearLayout;
import android.widget.TextView;
```

```java
public class MyAdapter extends BaseAdapter {
    private BookItem[] books = {new BookItem("Java 程序设计", R.drawable.png0001),
                    new BookItem("Android 应用开发", R.drawable.png0002),
                    new BookItem("Oracle 数据库管理指南", R.drawable.png0003),
                    new BookItem("Java Web 程序设计", R.drawable.png0004),
                    new BookItem("软件工程之系统工程师之路", R.drawable.png0005)};

    LayoutInflater inflater;
    int id_list_item;

    public MyAdapter(Context context, int id_list_item) {
        this.id_list_item = id_list_item;
        inflater = (LayoutInflater)
                    context.getSystemService(Context.LAYOUT_INFLATER_SERVICE);
    }

    @Override
    public int getCount() {
        return books.length;
    }

    @Override
    public Object getItem(int position) {
        return books[position];
    }

    @Override
    public long getItemId(int position) {
        return position;
    }

    @SuppressLint("ViewHolder") @Override
    public View getView(int position, View convertView, ViewGroup parent) {
        LinearLayout ll = (LinearLayout)inflater.inflate(id_list_item, parent, false);
        ImageView iv = (ImageView)ll.findViewById(R.id.id_book_photo);
        iv.setImageResource(books[position].photo);
        TextView tv;
        tv = (TextView)ll.findViewById(R.id.id_book_name);
        tv.setText(books[position].name);

        return ll;
    }

    private class BookItem {
        String name;
        int photo;

        public BookItem(String name, int photo) {
            this.name = name;
            this.photo = photo;
        }
    }
}
```

为了表示书的定义,我们定义了一个内部类 BookItem,其中包含书的名称和书的封面,同时，在 books 变量中初始化书的信息。为了能够显示图片，我们在 res/drawable 中放置了书的封面。

在 MyAdapter 的 getView()方法中，我们仍然采用相似的过程来封装要显示在 ListView 中的列表项：展开列表项布局资源，并在相应的 ImageView 组件中显示书的封面，在 TextView 组件中显示书的名称，并将这些封装好的组件返回给 ListView。

运行修改后的程序，将显示如图 5-15 所示的界面。

5.6.3.3 响应对 ListView 的列表项的点击

继续完善上面的例子，我们希望当用户点击列表中的某项时，在一个提示信息框中显示书的名称，例如，当点击"Android 应用开发"列表项时，显示如图 5-16 所示的信息。

图 5-16 使用 Toast 组件显示信息

为此，需要修改 MainActivity.java 文件，在其 onCreate()方法中设置 ListView 对 OnItemClickListener 事件的监听，并且实现 onItemClick()方法，修改后的 MainActivity.java 文件如下：

```java
package com.ttt.ex05listview01;

import android.os.Bundle;
import android.support.v7.app.AppCompatActivity;
import android.view.View;
import android.widget.AdapterView;
import android.widget.ListView;
import android.widget.TextView;
import android.widget.Toast;

public class MainActivity extends AppCompatActivity implements AdapterView.OnItemClickListener {

    @Override
    protected void onCreate(Bundle savedInstanceState) {
        super.onCreate(savedInstanceState);
        setContentView(R.layout.activity_main);

        ListView lv = (ListView)this.findViewById(R.id.id_lv);      //获得 ListView 组件
        MyAdapter mad = new MyAdapter(this, R.layout.list_item);
        lv.setAdapter(mad);      //告诉 ListView 从 mad 这个 Adapter 中获得数据
        lv.setOnItemClickListener(this);
    }

    @Override
    public void onItemClick(AdapterView<?> parent, View view, int position, long id) {
        TextView tv = (TextView)view.findViewById(R.id.id_book_name);
        String name = (String) tv.getText();
        Toast.makeText(this, name, Toast.LENGTH_LONG).show();
    }
}
```

我们只是对 MainActivity 类实现了 OnItemClickListener 接口，内容如下：

```
public class MainActivity extends AppCompatActivity implements AdapterView.
OnItemClickListener {
```

同时，在 MainActivity 的 onCreate()方法中设置 ListView 对象监听 OnItemClickListener，内容如下：

```
        lv.setOnItemClickListener(this);
```

并且，在 OnItemClickListener 接口的 onItemClick()方法中，显示被点击的列表项的信息，内容如下：

```
    @Override
    public void onItemClick(AdapterView<?> parent, View view, int position,
            long id) {
        TextView tv = (TextView)view.findViewById(R.id.id_book_name);
        String name = (String) tv.getText();
        Toast.makeText(this, name, Toast.LENGTH_LONG).show();
    }
```

注意 onItemClick()方法所传进来的参数，其中的 View view 就是被点击的对象，在例子程序中，它就是一个 LinearLayout，因此，我们可以从中找到显示书的名称的 TextView，并从这个 TextView 中得到书的名称。

Toast 是一个经常用来显示信息的 Android 组件，可以使用 Toast 提供的 make 静态方法来构造所要显示的信息，并调用 Toast 的 show()方法将所构建的 Toast 对象显示出来。

5.6.3.4 使用 ListView 显示 Android 平台已安装的所有程序

现在回到在 5.6.1 节开始提到显示 Android 平台中已经安装的所有应用程序的图标、名称和入口 Activity 的例子，现在来完成这个例子程序。首先在 Android Studio 中新建一个名为 Ex05ListViewEnd 的 Android 工程。

修改 activity_main.xml 主布局文件，内容如下：

```
<?xml version="1.0" encoding="utf-8"?>
<LinearLayout xmlns:android="http://schemas.android.com/apk/res/android"
    android:layout_width="match_parent"
    android:layout_height="match_parent">

    <ListView
        android:id="@+id/lv"
        android:layout_width="fill_parent"
        android:layout_height="wrap_content"/>

</LinearLayout>
```

就是在一个 LinearLayout 容器中包含一个 ListView 组件。

修改 list_item.xml 列表项布局文件，修改后的内容如下：

```
<?xml version="1.0" encoding="utf-8"?>
<LinearLayout xmlns:android="http://schemas.android.com/apk/res/android"
    android:layout_width="match_parent"
    android:layout_height="match_parent"
    android:orientation="horizontal" >

    <ImageView
        android:id="@+id/id_icon"
        android:layout_width="48dp"
```

```xml
        android:layout_height="48dp"
        android:contentDescription="@string/text_empty" />

    <LinearLayout
        android:layout_width="match_parent"
        android:layout_height="wrap_content"
        android:orientation="vertical" >

        <TextView
            android:id="@+id/id_appName"
            android:layout_width="match_parent"
            android:layout_height="wrap_content" />

        <TextView
            android:id="@+id/id_packageName"
            android:layout_width="match_parent"
            android:layout_height="wrap_content" />
    </LinearLayout>

</LinearLayout>
```

修改 MyAdapter.java 文件,这个 Adapter 将从 Android 平台获得所有已安装的应用,并创建列表项提交给 ListView 显示,修改后的内容如下:

```java
package com.ttt.Ex05ListViewEnd;

import java.util.ArrayList;
import java.util.HashMap;
import java.util.List;
import java.util.Map;

import com.ttt.ex05listview01.R;

import android.annotation.SuppressLint;
import android.content.Context;
import android.content.pm.PackageInfo;
import android.content.pm.PackageManager;
import android.graphics.drawable.Drawable;
import android.view.LayoutInflater;
import android.view.View;
import android.view.ViewGroup;
import android.widget.BaseAdapter;
import android.widget.ImageView;
import android.widget.TextView;

public class MyAdapter extends BaseAdapter {
    ArrayList<HashMap<String, Object>> items =
                            new ArrayList<HashMap<String, Object>>();
    private LayoutInflater mInflater;
    int item_layout;

    public MyAdapter(Context context, int item_layout) {
        mInflater = (LayoutInflater) context
                .getSystemService(Context.LAYOUT_INFLATER_SERVICE);
        this.item_layout = item_layout;
```

```java
        PackageManager pm = context.getPackageManager(); // 得到 PackageManager 对象
        // 得到系统安装的所有程序包的 PackageInfo 对象
        List<PackageInfo> packs = pm.getInstalledPackages(0);
        for (PackageInfo pi : packs) {
            HashMap<String, Object> map = new HashMap<String, Object>();
            map.put("icon", pi.applicationInfo.loadIcon(pm));        // 图标
            map.put("appName", pi.applicationInfo.loadLabel(pm));// 应用名
            map.put("packageName", pi.packageName);                  // 包名
            items.add(map);                                          // 循环读取存到 HashMap
        }
    }

    @Override
    public int getCount() {
        return items.size();
    }

    @Override
    public Object getItem(int position) {
        return items.get(position);
    }

    @Override
    public long getItemId(int position) {
        return position;
    }

    @SuppressLint("ViewHolder")
    public View getView(int position, View convertView, ViewGroup parent) {
        View v;
        v = mInflater.inflate(item_layout, parent, false);

        final Map<String, ?> dataSet = items.get(position);
        if (dataSet == null) {
            return null;
        }

        ImageView iv = (ImageView)v.findViewById(R.id.id_icon);
        iv.setImageDrawable((Drawable) dataSet.get("icon"));
        TextView tv1 = (TextView)v.findViewById(R.id.id_appName);
        tv1.setText(dataSet.get("appName").toString());
        TextView tv2 = (TextView)v.findViewById(R.id.id_packageName);
        tv2.setText(dataSet.get("packageName").toString());

        return v;
    }
}
```

在 MyAdapter 的构造函数中，我们首先获得一个布局展开器对象，用于在 getView()方法中展开列表项布局，然后获得一个包管理器对象，从 Android 平台中获得所有已安装的应用信息，并将每个应用的信息作为一个 Map 对象存入到 ArrayList 对象中。在 getView()方法中，我们首先展开一个列表项布局，从 items 数组中获得相应的应用信息，然后逐个置入到展开的列表项布局中，最后将包装好的列表项组件返回给 ListView。

修改 MainActivity.java 程序为如下内容：

```java
package com.ttt. Ex05ListViewEnd;

import android.os.Bundle;
import android.support.v7.app.AppCompatActivity;
import android.view.View;
import android.widget.AdapterView;
import android.widget.ListView;
import android.widget.TextView;
import android.widget.Toast;

public class MainActivity extends AppCompatActivity {

    @Override
    protected void onCreate(Bundle savedInstanceState) {
        super.onCreate(savedInstanceState);
        setContentView(R.layout.activity_main);

        ListView lv = (ListView)this.findViewById(R.id.lv);    //获得ListView组件
        MyAdapter mad = new MyAdapter(this, R.layout.list_item);
        lv.setAdapter(mad);    //告诉ListView从mad这个Adapter中获得数据

    }

}
```

5.6.4 Spinner

Spinner 是我们所熟悉的下拉列表框。与 ListView 类似，我们必须为 Spinner 对象指定一个 Adapter。我们从 Spinner 的简单用法开始讲述。

5.6.4.1 Spinner 的简单应用

我们将构建一个简单的应用：在应用界面上显示一个简单的下拉列表，用户可以从中选择一个选项，并将该选项在界面上显示出来。程序的运行效果如图 5-17 所示。

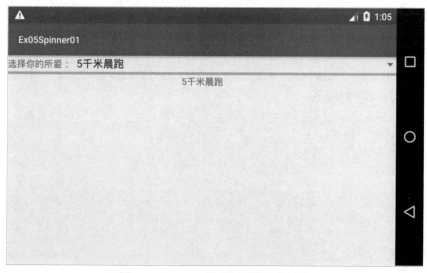

图 5-17　Spinner 简单例子的运行效果

选中其中一个选项,则在下拉列表下面的一个 TextView 中显示所选选项。为此,在 Android Studio 中新建一个名为 Ex05Spinner01 的 Android 工程。

首先修改主界面的布局文件,也就是 res/layout/content_main.xml 文件,修改后的内容如下:

```xml
<?xml version="1.0" encoding="utf-8"?>
<LinearLayout xmlns:android="http://schemas.android.com/apk/res/android"
    android:layout_width="match_parent"
    android:layout_height="match_parent"
    android:orientation="vertical">

    <LinearLayout
        android:layout_width="match_parent"
        android:layout_height="wrap_content"
        android:orientation="horizontal" >

        <TextView
            android:layout_width="wrap_content"
            android:layout_height="wrap_content"
            android:text="@string/text_choice" />
        <Spinner
            android:id="@+id/id_spinner"
            android:layout_width="match_parent"
            android:layout_height="wrap_content" />

    </LinearLayout>

    <View
        android:layout_width="match_parent"
        android:layout_height="4dp"
        android:background="#aaa" />

    <TextView
        android:id="@+id/id_choice"
        android:gravity="center"
        android:layout_width="match_parent"
        android:layout_height="wrap_content" />

</LinearLayout>
```

这个布局比较简单,就是在 LinearLayout 中放置一个 LinearLayout、一个分隔组件用的 View 和一个显示所选结果的 TextView。在其中嵌套的 LinearLayout 中放置了一个用于提示用的 TextView 和下拉列表框 Spinner。

现在修改 res/values/strings.xml 文件,首先定义在布局文件中引用到的资源,然后,定义一个字符串数组资源,修改后的 strings.xml 文件内容如下:

```xml
<?xml version="1.0" encoding="utf-8"?>
<resources>

    <string name="app_name">Ex05Spinner01</string>
    <string name="action_settings">Settings</string>

    <string name="text_choice">选择你的所爱:</string>

    <string-array name="habit">
        <item>5千米晨跑</item>
```

```
        <item>去健身房健身</item>
        <item>爬山</item>
        <item>2 千米游泳</item>
        <item>打篮球</item>
    </string-array>

</resources>
```

在代码中我们定义了一个字符串数组资源,所定义的这个字符串数组资源正是下拉选项,接下来将在 MainActivity.java 中介绍如何使用这个字符串数组资源。

最后,修改 MainActivity.java 文件,修改后的内容如下:

```java
package com.ttt.ex05spinner01;

import android.os.Bundle;
import android.support.v7.app.AppCompatActivity;
import android.view.View;
import android.widget.AdapterView;
import android.widget.ArrayAdapter;
import android.widget.Spinner;
import android.widget.TextView;

public class MainActivity extends AppCompatActivity  implements AdapterView.OnItemSelectedListener {
    TextView choice;
    ArrayAdapter<CharSequence> adapter;

    @Override
    protected void onCreate(Bundle savedInstanceState) {
        super.onCreate(savedInstanceState);
        setContentView(R.layout.activity_main);

        choice = (TextView)this.findViewById(R.id.id_choice);

        Spinner spinner = (Spinner)this.findViewById(R.id.id_spinner);
        adapter = ArrayAdapter.createFromResource(this,
                R.array.habit, android.R.layout.simple_spinner_item);
        adapter.setDropDownViewResource(
                android.R.layout.simple_spinner_dropdown_item);
        spinner.setAdapter(adapter);
        spinner.setOnItemSelectedListener(this);
    }

    @Override
    public void onItemSelected(AdapterView<?> parent, View view, int position,
                          long id) {
        choice.setText(adapter.getItem(position));
    }

    @Override
    public void onNothingSelected(AdapterView<?> parent) {
    }
}
```

在 MainActivity 的 onCreate()方法中,我们首先获得 Spinner 对象,然后使用 ArrayAdapter

的静态方法 createFromResource()来创建一个 ArrayAdapter 对象,其中的 R.array.habit 参数用于指定一个作为字符串数组的资源,android.R.layout.simple_spinner_item 参数用于指定在下拉列表框未下拉时的数据显示布局,这里,我们使用了 Android 平台自带的一个布局,这个布局中只有一个 TextView。注意其中的语句:

```
adapter.setDropDownViewResource(
            android.R.layout.simple_spinner_dropdown_item);
```

在这里又指定了一个布局,这个布局指定的是当下拉列表展开时的布局,我们仍然使用 Android 平台自带的一个布局,这个布局仍然只有一个 TextView。最后,将构建好的 adapter 设置为 Spinner 的适配器。

为了响应对下拉列表项的选择事件,MainActivity 实现了 OnItemSelectedListener 接口,并在其中的 onItemSelected()方法中将 adapter 中对应的信息显示在第二个 TextView 中。

运行这个程序,并选中其中的一个选项,将得到如图 5-17 所示的结果。

5.6.4.2 美化 Spinner

关于上一个 Spinner 的例子,在显示下拉列表框时显得太单调了。在这一节,我们要美化这个例子,使之显示下拉列表时更加美观。修改后的程序的运行效果如图 5-18 所示。

图 5-18 修改后的 Spinner 下拉列表

为此,需要修改工程 Ex05Spinner01,添加一个 Java 文件、几个图片资源和一个下拉列表布局文件。

首先添加几个用于显示在下拉列表中的图片资源,然后,创建一个新的布局资源 dropdown_item.xml,这个布局资源就是下拉列表在下拉时的布局,其内容如下:

```xml
<?xml version="1.0" encoding="utf-8"?>
<LinearLayout xmlns:android="http://schemas.android.com/apk/res/android"
    android:layout_width="match_parent"
    android:layout_height="match_parent"
    android:orientation="horizontal" >

    <ImageView
        android:id="@+id/id_imageview"
        android:layout_width="64dp"
        android:layout_height="64dp" />
```

```xml
<LinearLayout
    android:layout_width="match_parent"
    android:layout_height="match_parent"
    android:orientation="vertical" >

    <TextView
        android:id="@+id/id_textview_01"
        android:textSize="16sp"
        android:layout_width="match_parent"
        android:layout_height="wrap_content" />

    <TextView
        android:id="@+id/id_textview_02"
        android:textSize="12sp"
        android:layout_width="match_parent"
        android:layout_height="wrap_content" />

</LinearLayout>
</LinearLayout>
```

这个布局文件很简单,就是对如图 5-18 所示的下拉列表项的布局。

为了选择在下拉列表时显示我们自定义的效果,需要重新定义 ArrayAdapter,因此,我们创建了一个自定义的 **MyArrayAdapter**,并重写了其中的 **getDropDownView()** 方法,注意这个方法是在用户选择下拉列表项时被调用的,因此,我们可以在这个方法中创建指定的布局,并显示指定的信息。**MyArrayAdapter.java** 文件的内容如下:

```java
package com.ttt.ex05spinner01;

import android.content.Context;
import android.view.LayoutInflater;
import android.view.View;
import android.view.ViewGroup;
import android.widget.ArrayAdapter;
import android.widget.ImageView;
import android.widget.TextView;

public class MyArrayAdapter extends ArrayAdapter<CharSequence> {
    private LayoutInflater mInflater;
    String[] titles;
    String[] desc = {
                    "所谓5千米晨跑,就是不能少于5千米,当然,你可以多跑",
                    "所谓去健身房健身,就是要亲自练,而不是看别人练",
                    "所谓爬山,山的高度起码要有1000米,两个来回",
                    "所谓2千米游泳,这可有点厉害",
                    "所谓打篮球,是指10个人比赛,不是单打独斗"
                };
    int[] images = {R.drawable.png0010, R.drawable.png0011,
            R.drawable.png0012, R.drawable.png0013,
            R.drawable.png0014};

    public MyArrayAdapter(Context context, int resource, CharSequence[] objects) {
        super(context, resource, objects);

        mInflater = (LayoutInflater) context
```

```
            .getSystemService(Context.LAYOUT_INFLATER_SERVICE);
        titles = context.getResources().getStringArray(R.array.habit);
    }

    @Override
    public int getCount() {
        return titles.length;
    }

    @Override
    public View getDropDownView(int position, View convertView, ViewGroup parent) {
        View v;
        v = mInflater.inflate(R.layout.dropdown_item, parent, false);

        ImageView iv = (ImageView)v.findViewById(R.id.id_imageview);
        iv.setImageResource(images[position]);
        TextView tv01 = (TextView)v.findViewById(R.id.id_textview_01);
        tv01.setText(titles[position]);
        TextView tv02 = (TextView)v.findViewById(R.id.id_textview_02);
        tv02.setText(desc[position]);

        return v;
    }
}
```

在这个 Adapter 的构造函数中，我们使用语句：

```
        titles = context.getResources().getStringArray(R.array.habit);
```

这条语句，是指从资源中获得定义的字符串数组资源，也就是说，在资源文件中定义的字符串数组资源，可以很便利地在 Java 代码中引用它。然后，在 Adapter 的 getDropDownView() 方法中，我们展开下拉列表项布局，并将相应的信息写入到相应的组件中，最后将组装好的组件返回给 Spinner，以便在选择下拉列表时显示出来。

然后需要修改 MainActivity.java 文件，修改后的文件内容如下：

```
package com.ttt.ex05spinner01;

import android.os.Bundle;
import android.view.View;
import android.view.MenuItem;
import android.widget.AdapterView;
import android.widget.ArrayAdapter;
import android.widget.Spinner;
import android.widget.TextView;

public class MainActivity extends AppCompatActivity implements AdapterView.OnItemSelectedListener {
    TextView choice;
    ArrayAdapter<CharSequence> adapter;

    @Override
    protected void onCreate(Bundle savedInstanceState) {
        super.onCreate(savedInstanceState);
        setContentView(R.layout.activity_main);
```

```
    choice = (TextView)this.findViewById(R.id.id_choice);

    Spinner spinner = (Spinner)this.findViewById(R.id.id_spinner);
    adapter = new MyArrayAdapter(this, android.R.layout.simple_spinner_item,
         this.getResources().getTextArray(R.array.habit));
    spinner.setAdapter(adapter);

    spinner.setOnItemSelectedListener(this);
}

@Override
public void onItemSelected(AdapterView<?> parent, View view, int position,
                   long id) {
    choice.setText(adapter.getItem(position));
}

@Override
public void onNothingSelected(AdapterView<?> parent) {
}
}
```

在 onCreate()方法中我们创建了自定义的 MyArrayAdapter 对象，并将它设置为 Spinner 的适配器，如此而已。

运行修改后的程序，将得到如图 5-18 所示的效果。

5.6.5　GridView

GridView 以二维表格的方式显示数据，若数据比较多，该组件将提供垂直滚动条。我们用一个例子来说明 GridView 的使用方法：在这个例子中，我们用 GridView 显示一组图片的缩略图，当点击某个缩略图时，用 Toast 组件显示这个图片的信息，运行效果如图 5-19 所示。

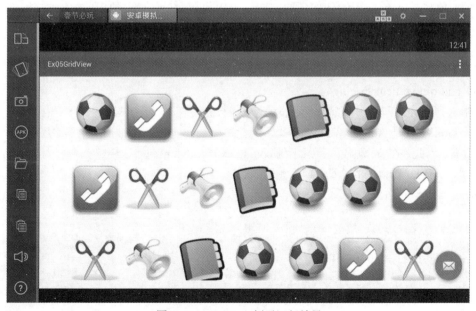

图 5-19　GridView 例子运行效果

为此，我们在 Android Studio 中建立一个名称为 Ex05GridView 的工程，并在 res/drawable 目录下放置一些需要显示的图片资源。

首先修改用于显示该界面的布局文件 res/content_main.xml 文件，修改后的文件内容如下：

```xml
<?xml version="1.0" encoding="utf-8"?>
<RelativeLayout xmlns:android="http://schemas.android.com/apk/res/android"
    android:layout_width="match_parent"
    android:layout_height="match_parent">

    <GridView
        android:id="@+id/gridview"
        android:layout_width="match_parent"
        android:layout_height="match_parent"
        android:gravity="center"
        android:columnWidth="90dp"
        android:numColumns="auto_fit"
        android:stretchMode="columnWidth"
        android:horizontalSpacing="10dp"
        android:verticalSpacing="10dp" />

</RelativeLayout>
```

这个布局比较简单，就是在一个 RelativeLayout 布局中放置一个 GridView。GridView 的关键配置属性如下。

（1）android:columnWidth。

其指定 GridView 每列的宽度，需要设定一个宽度值。

（2）android:numColumns。

其指定每行的列数，取值可以为一个整数值，也可以为"auto_fit"，即自动适应。

（3）android:stretchMode。

其指定当屏幕的宽度值不是 android:columnWidth 的整数倍时，多余的显示空间的分配方式。它的取值包括 none，不做任何延展；spacingWidth，将列之间的间隔放大；columnWidth，将每列自动放大；spacingWidthUniform，将多余的空间在列之间均匀分配。

（4）android:horizontalSpacing。

其指定行与行之间的分隔高度值，取值为一个整数值。

（5）android:verticalSpacing。

其指定列与列之间的分隔宽度值，取值为一个整数值。

由于 GridView 是 AdapterView 的子类，因此，需要创建一个 Adapter 的接口类来为它提供要显示的数据。为此，在 src 源代码目录下的 com.ttt.ex05gridview 包下，新建一个名为 ImageAdapter 的 Java 类，ImageAdapter 的代码如下：

```java
package com.ttt.ex05gridview;

import com.example.ex05gridview.R;

import android.content.Context;
import android.view.View;
import android.view.ViewGroup;
import android.widget.BaseAdapter;
import android.widget.GridView;
import android.widget.ImageView;
```

```java
public class ImageAdapter extends BaseAdapter {
    private Context mContext;

    public ImageAdapter(Context c) {
        mContext = c;
    }

    public int getCount() {
        return mThumbIds.length;
    }

    public Object getItem(int position) {
        return null;
    }

    public long getItemId(int position) {
        return 0;
    }

    public View getView(int position, View convertView, ViewGroup parent) {
        ImageView imageView;
        if (convertView == null) {
            imageView = new ImageView(mContext);
            int width = GridView.LayoutParams.MATCH_PARENT;
            int height = GridView.LayoutParams.MATCH_PARENT;
            imageView.setLayoutParams(new GridView.LayoutParams(width, height));
            imageView.setScaleType(ImageView.ScaleType.CENTER_INSIDE);
        } else {
            imageView = (ImageView) convertView;
        }

        imageView.setImageResource(mThumbIds[position]);
        return imageView;
    }

    private Integer[] mThumbIds = {
            R.drawable.png0029, R.drawable.png0030,
            R.drawable.png0031, R.drawable.png0032,
            R.drawable.png0033, R.drawable.png0029,
            R.drawable.png0029, R.drawable.png0030,
            R.drawable.png0031, R.drawable.png0032,
            R.drawable.png0033, R.drawable.png0029,
            R.drawable.png0029, R.drawable.png0030,
            R.drawable.png0031, R.drawable.png0032,
            R.drawable.png0033, R.drawable.png0029,
            R.drawable.png0029, R.drawable.png0030,
            R.drawable.png0031, R.drawable.png0032,
            R.drawable.png0033, R.drawable.png0029
    };
}
```

创建自己的 Adapter，并将需要显示的图片资源保存在 mThumbIds 数组中。注意这里的 getView() 方法的实现：

```java
public View getView(int position, View convertView, ViewGroup parent) {
    ImageView imageView;
```

```
        if (convertView == null) {
            imageView = new ImageView(mContext);
            int width = GridView.LayoutParams.MATCH_PARENT;
            int height = GridView.LayoutParams.MATCH_PARENT;
            imageView.setLayoutParams(new GridView.LayoutParams(width, height));
            imageView.setScaleType(ImageView.ScaleType.CENTER_INSIDE);
        } else {
            imageView = (ImageView) convertView;
        }

        imageView.setImageResource(mThumbIds[position]);
        return imageView;
    }
```

需要特别解释的是参数 View convertView 的使用。convertView 是可重用的对象,它是呈现 GridView 视图的时候返回给 GridView 的 ImageView 对象,但是现在由于用户滚动屏幕,所以这个对象即将从屏幕上消失,我们知道,Java 虚拟机在创建和销毁对象时是比较花费时间的,为了提高程序的运行效率,Android 将这个即将被销毁的对象返回给 getView()方法,判断是否可以重用这个对象,以便提升系统的效率。在 getView()方法中,正是利用了这个特点:先判断这个 convertView 的值是否为空,如果不为空,那么,我们可以重用这个对象进而减少重新创建一个 ImageView 对象的时间。

在这段代码中,我们还对即将显示在 GridView 中的 ImageView 进行布局:通过代码指定 ImageView 在水平和垂直方向上均占满 GridView 为 ImageView 分配的显示空间。

现在修改 MainActivity.java 代码,onCreate()方法用于获得界面上的 GridView 的引用,设置 GridView 的监听事件,并用 Toast 组件显示被点击的图片的序号。MainActivity.java 代码如下:

```
package com.ttt.ex05gridview;

import android.os.Bundle;
import android.support.v7.app.AppCompatActivity;
import android.view.View;
import android.widget.AdapterView;
import android.widget.GridView;
import android.widget.Toast;

public class MainActivity extends AppCompatActivity {

    @Override
    protected void onCreate(Bundle savedInstanceState) {
        super.onCreate(savedInstanceState);
        setContentView(R.layout.activity_main);

        GridView gridview = (GridView) findViewById(R.id.gridview);
        gridview.setAdapter(new ImageAdapter(this));
        gridview.setOnItemClickListener(new AdapterView.OnItemClickListener() {
            @Override
            public void onItemClick(AdapterView<?> parent, View view, int position, long id) {
                Toast.makeText(MainActivity.this, "" + position, Toast.LENGTH_SHORT).show();
            }
```

```
        });
    }
}
```

运行修改完成的程序，即可显示如图 5-19 所示的界面。当我们点击某个图片时，将在 Toast 组件中显示被点击的图片的序号。

5.7 同步练习三

1. 完善 5.6.3.2 节的例子，使之不仅可以显示书的封面、名称，还可以同时显示书的作者及书的出版社名称。

2. Spinner 探究：读者可以从任何途径查询资料，通过自我探究来修改 Spinner 下拉列表展开时的显示样式。

5.8 Android 其他常用组件

在 Android 平台的 SDK 中，android.widget 包中还有许多组件，在此，我们不再一一介绍，表 5-4 给出了 Android 常用的组件名称及其含义。

表 5-4 Android 常用的组件名称及其含义

组 件 名 称	含 义
CalendarView	这个组件用于显示日历，用户可以从中选择日期
Chronometer	一个简单的计时器组件
DatePicker	日期选择组件
HorizontalScrollView	带有水平滚动条的布局容器组件
ImageButton	图片按钮组件
NumberPicker	允许用户从一个指定范围选择一个整数
ProgressBar	进度条组件
Space	在布局中可用于分隔组件以使组件之间留下间隙
TabHost	多标签窗口组件
TableLayout	表格布局容器
TableRow	与 TableLayout 配合使用的表格布局中的一行
TextClock	用指定的格式显示当前的日期和时间
TimePicker	时间选择组件
ViewFlipper	以动画的方式切换组件的显示
WebView	用于显示 Web 页面的组件

这个列表不完整，完整的内容请读者参考 Android 的帮助文档。

5.9 同步练习四

自我学习：编写一个程序，使其可以用 WebView 组件来浏览网页。关于 WebView 组件的使用方法，读者可以参考 Android 的帮助文档。

样式和主题

在进行程序的界面设计时,我们经常需要对界面及界面上的组件设置统一的显示外观,例如,界面的背景颜色、使用的字体大小、字体颜色、组件的显示大小、是否显示标题栏等,我们可以为每个组件设置自己的显示属性,但是,为了便于对外观的统一管理,需要将这些外观设置集中起来。Android 是通过样式,也就是 Style 来完成这个工作的。想要在 Android 中使用样式来定制外观,需要做两方面的工作:定义样式;将定义好的样式应用到界面中。先从一个简单的例子程序开始讲述。

6.1 样式入门

我们先通过一个简单的例子来看一下 Android 是如何定义样式和如何将定义好的样式应用到界面中的。为此,新建一个名称为 Ex06StyleTheme01 的 Android 工程,然后修改 res/layout/activity_main.xml 文件,内容如下:

```xml
<?xml version="1.0" encoding="utf-8"?>
<LinearLayout xmlns:android="http://schemas.android.com/apk/res/android"
    xmlns:app="http://schemas.android.com/apk/res-auto"
    xmlns:tools="http://schemas.android.com/tools"
    android:layout_width="match_parent"
    android:layout_height="match_parent"
    android:orientation="vertical"
    android:paddingBottom="@dimen/activity_vertical_margin"
    android:paddingLeft="@dimen/activity_horizontal_margin"
    android:paddingRight="@dimen/activity_horizontal_margin"
    android:paddingTop="@dimen/activity_vertical_margin"
    app:layout_behavior="@string/appbar_scrolling_view_behavior"
    tools:context="com.ttt.ex06styletheme01.MainActivity"
    tools:showIn="@layout/activity_main">

    <Button
        android:layout_width="match_parent"
        android:layout_height="0dp"
        android:layout_weight="1"
        android:text="@string/text_btn_01" />

    <Button
        android:layout_width="match_parent"
        android:layout_height="0dp"
```

```
        android:layout_weight="1"
        android:text="@string/text_btn_02" />

    <TextView
        android:layout_width="match_parent"
        android:layout_height="0dp"
        android:layout_weight="1"
        android:gravity="center"
        android:text="@string/text_textview" />

</LinearLayout>
```

这个布局文件很简单：就是在一个 LinearLayout 容器中包含两个按钮和一个文本框，并且这 3 个组件平分 LinearLayout 的显示空间。继续修改 res/values/strings.xml 文件，在其中定义几个字符串常量资源，内容如下：

```
<?xml version="1.0" encoding="utf-8"?>
<resources>

    <string name="app_name">Ex06StyleTheme01</string>
    <string name="action_settings">Settings</string>

    <string name="text_btn_01">第一个按钮</string>
    <string name="text_btn_02">第二个按钮</string>
    <string name="text_textview">学好 Android 的样式和主题</string>

</resources>
```

代码中定义的 3 个常量资源，其实就是在布局文件中引用到的 3 个字符串常量。现在运行该程序，结果如图 6-1 所示。

图 6-1　Ex06StyleTheme01 运行结果

从图 6-1 可以看出，两个按钮和一个文本框都使用了 Android 自定义的默认样式，现在我们来修改按钮及文本框的样式。在 res/values 工程目录下，新建一个名为 mystyles.xml 的文件，并修改其内容为如下代码：

```xml
<resources>

    <style name="MyButtonStyle">
        <item name="android:textColor">#00FF00</item><!--绿色-->
        <item name="android:background">#FF0000</item><!--红色-->
        <item name="android:textSize">16sp</item>
    </style>

    <style name="MyTextViewStyle">
        <item name="android:textColor">#0000FF</item><!--黑色-->
        <item name="android:typeface">monospace</item>
    </style>

</resources>
```

在 res/values/mystyles.xml 文件中，我们定义了两个新的 style 样式：其中一个 style 的 name 为 MyButtonStyle，另一个为 MyTextViewStyle。在 MyButtonStyle 的 style 中，我们定义了文本颜色、背景和文字大小；在 MyTextViewStyle 的 style 中，我们定义了字体颜色和字体类型。

现在将这两个已经定义好的 style 应用到界面的组件中：将 MyButtonStyle 应用到界面的第一个 Button 组件中，将 MyTextViewStyle 应用到界面的 TextView 组件中。为此，修改 res/layout/activity_main.xml 文件，内容如下：

```xml
<?xml version="1.0" encoding="utf-8"?>
<LinearLayout xmlns:android="http://schemas.android.com/apk/res/android"
    xmlns:app="http://schemas.android.com/apk/res-auto"
    xmlns:tools="http://schemas.android.com/tools"
    android:layout_width="match_parent"
    android:layout_height="match_parent"
    android:orientation="vertical"
    android:paddingBottom="@dimen/activity_vertical_margin"
    android:paddingLeft="@dimen/activity_horizontal_margin"
    android:paddingRight="@dimen/activity_horizontal_margin"
    android:paddingTop="@dimen/activity_vertical_margin"
    app:layout_behavior="@string/appbar_scrolling_view_behavior"
    tools:context="com.ttt.ex06styletheme01.MainActivity"
    tools:showIn="@layout/activity_main">

    <Button
        android:layout_width="match_parent"
        android:layout_height="0dp"
        android:layout_weight="1"
        style="@style/MyButtonStyle"      //将定义好的style应用到这个组件中
        android:text="@string/text_btn_01" />

    <Button
        android:layout_width="match_parent"
        android:layout_height="0dp"
        android:layout_weight="1"
```

```
        android:text="@string/text_btn_02" />

    <TextView
        android:layout_width="match_parent"
        android:layout_height="0dp"
        android:layout_weight="1"
        android:gravity="center"
        style="@style/MyTextViewStyle"    //将定义好的style应用到这个组件中
        android:text="@string/text_textview" />

</LinearLayout>
```

注意第一个按钮中的代码:

```
style="@style/MyButtonStyle"        //将定义好的style应用到这个组件中
```

这句代码表示将定义好的 MyButtonStyle 应用到这个 Button 组件中。同时，注意 TextView 中的代码:

```
style="@style/MyTextViewStyle"     //将定义好的style应用到这个组件中
```

这句代码表示将定义好的 MyTextViewStyle 应用到这个 TextView 组件中。运行修改后的程序，效果如图 6-2 所示。

图 6-2　添加样式后的 Ex06StyleTheme01 程序的运行效果

比较图 6-1 和图 6-2 可以看出添加样式前后的程序运行效果的不同。通过这个例子，我们对 Android 的样式的定义和使用方法有了一个初步的了解。下面笔者将详细介绍如何定义样式和如何使用样式。

6.2 定义样式

6.2.1 定义样式的一般方法

为了定义一个样式，需要在 res/values 工程目录下新建一个 XML 文件，当然，也可以在现有的某个文件中（如 styles.xml 文件）直接添加要定义的样式。定义样式的一般格式如下：

```xml
<?xml version="1.0" encoding="utf-8"?>
<resources>

    <style name="自定义样式名称" parent="父样式名称">
        <item name="样式属性名称">属性值</item>
        ……
    </style>

    <style name="自定义样式名称" parent="父样式名称">
        <item name="样式属性名称">属性值</item>
        ……
    </style>

    ……

</resources>
```

可以在 Java 程序中使用 "R.style.定义样式名称" 来访问所定义的样式，也可以在 XML 文件中使用 "@style/自定义样式名称" 来访问。注意，在样式定义中 parent="父样式名称"，这意味着样式定义是支持继承的，也就是我们常说的级联样式，同时，样式定义中的 parent 属性是可选的。我们来看一个样式定义的例子，内容如下：

```xml
<?xml version="1.0" encoding="utf-8"?>
<resources>

    <style name="GreenText" parent="@android:style/TextAppearance">
        <item name="android:textColor">#00FF00</item>
    </style>

</resources>
```

在这个样式定义中，我们定义了一个新的名称为 GreenText 的样式，它继承了 Android 平台已定义的名称为 "@android:style/TextAppearance" 的样式，同时，修改其中的 android:textColor 属性的值为 "#00FF00"，也就是绿色。上面我们定义了名为 GreenText 的样式，现在，我们可以通过继承来定义新的样式，例如，定义一个名为 GreenTextLarge 的样式，内容如下：

```xml
<?xml version="1.0" encoding="utf-8"?>
<resources>

    <style name="GreenText" parent="@android:style/TextAppearance">
        <item name="android:textColor">#00FF00</item>
    </style>

    <style name="GreenTextLarge" parent="@style/GreenText">
```

```xml
        <item name="android:textSize">32sp</item>
    </style>

</resources>
```

在这里，我们采用继承的方式定义了一个新的名为 GreenTextLarge 的样式。由于 GreenText 样式是我们自定义的样式，所以还可以使用如下的方式来定义样式的继承：

```xml
<?xml version="1.0" encoding="utf-8"?>
<resources>

    <style name="GreenText" parent="@android:style/TextAppearance">
        <item name="android:textColor">#00FF00</item>
    </style>

    <style name="GreenTextLarge" parent="@style/GreenText">
        <item name="android:textSize">32sp</item>
    </style>

    <style name="GreenText.Small">     //针对自定义的父样式，可以采用这种方式来继承
        <item name="android:textSize">8sp</item>
    </style>

</resources>
```

这里定义了一个新的名为 GreenText.Small 的样式，注意这是一个特殊的样式名称，这个名称表示 GreenText.Small 是一个新的样式，并且它的父样式是 GreenText 样式。

6.2.2 样式定义中的可用属性

样式定义中可以使用的属性随着定义样式的应用目标不同而不同，例如，如果定义一个针对 TextView 组件的样式和一个针对 Button 组件的样式，其中可以使用的属性是不同的。因此，在定义样式时，应该针对样式未来的应用目标，参考组件的可用 XML 属性来确定可用属性。有一种例外是，如果将某个样式定义应用到某个组件，而在这个样式定义中包含某个所应用到的组件不支持的属性，那么，这个组件会自动忽略掉这个不支持的属性，而不会影响其他支持的属性所起的作用。

如果读者对 Android 支持的完整属性列表感兴趣，可以参考 Android 帮助文档中的 android.R.styleable 这个类，在这个类中，针对每个 Android 组件，文档都给出了完整的样式定义可用属性。例如，在 android.R.styleable 这个类中，查看 ImageView 组件的部分可用属性，如图 6-3 所示。

```
public static final int[] ImageView
Attributes that can be used with a ImageView.
Includes the following attributes:
```

Attribute	Description
android:adjustViewBounds	Set this to true if you want the ImageView to adjust its bounds to preserve the ratio of its drawable.
android:baseline	The offset of the baseline within this view.
android:baselineAlignBottom	If true, the image view will be baseline aligned with based on its bottom edge
android:cropToPadding	If true, the image will be cropped to fit within its padding.
android:maxHeight	An optional argument to supply a maximum height for this view.
android:maxWidth	An optional argument to supply a maximum width for this view.
android:scaleType	Controls how the image should be resized or moved to match the size of this Ir
android:src	Sets a drawable as the content of this ImageView.

图 6-3 ImageView 组件的部分可用属性

同时，我们也知道，ImageView 是 View 的子类，因此，View 的可用属性也是 ImageView 的可用属性，在 android.R.styleable 类中，View 组件的部分可用属性如图 6-4 所示。

```
public static final int[] View
Attributes that can be used with View or any of its subclasses. Also see ViewGroup_Layout for attributes that processed by the view's parent.
Includes the following attributes:
```

Attribute	Description
android:accessibilityLiveRegion	Indicates to accessibility services whether the user should be when this view changes.
android:alpha	alpha property of the view, as a value between 0 (completely transparent) and 1 (completely opaque).
android:background	A drawable to use as the background.
android:backgroundTint	Tint to apply to the background.
android:backgroundTintMode	Blending mode used to apply the background tint.
android:clickable	Defines whether this view reacts to click events.
android:contentDescription	Defines text that briefly describes content of the view.

图 6-4 View 组件的部分可用属性

6.3 应用样式

一旦完成样式定义后，我们就可以将定义好的样式应用到需要的地方：可以将样式应用到某个组件中，也可以将样式应用到某个 Activity 或整个 Application 中。

6.3.1 将样式应用到某个组件

将定义好的样式应用到某个组件是一项非常简单的工作:在组件的配置中,添加styleXML配置属性即可。例如,我们要将上面定义的 GreenText.Small 样式应用到 TextView 的定义中,只需要添加 style 属性即可,内容如下:

```
<TextView
    style="@style/GreenText.Small"
    ……
    android:text="@string/hello" />
```

我们可以将样式应用到具体组件,也可以将样式应用到容器组件,注意,应用到容器组件的样式只对这个容器组件本身有效,而对放置于这个容器中的子组件是不起作用的。

6.3.2 将样式应用到某个 Activity 或整个 Application

本章的标题是"样式和主题",可是到目前为止我们还没有介绍什么是主题。那么,到底什么是主题呢?当我们把样式应用到某个 Activity 或整个 Application 的时候,这个样式就成为了主题。将样式应用到某个 Activity 或整个 Application,需要在 AndroidManifest.xml 文件中针对某个 Activity 或整个 Application 添加 android:theme 属性,例如,我们可以先定义如下的一个样式:

```
<?xml version="1.0" encoding="utf-8"?>
<resources>
    <color name="custom_theme_color">#b0b0ff</color>
    <style name="CustomTheme" parent="@style/MyTheme.Light">
        <item name="android:windowBackground">@color/custom_theme_color</item>
        <item name="android:colorBackground">@color/custom_theme_color</item>
    </style>
</ resources >
```

然后,将这个样式应用到某个 Activity,内容如下:

```
<activity android:theme="@style/CustomTheme">
```

或将这个样式应用到整个 Application,内容如下:

```
<application android:theme="@style/CustomTheme">
```

那么这个样式就是主题。

主题是一种特殊的样式,由于这种主题样式是应用于某个 Activity 或整个 Application 的,因此,Android 为主题样式的定义引入了一些特殊的属性,例如,android:windowNoTitle 属性就只能用于对主题样式的定义中,它表示当该属性被应用到某个 Activity 或 Application 中时,不显示 Activity 的标题;android:textSelectHandle 属性表示在进行文本选择时所显示的文本定位图片资源;等等。完整的可用于主题样式定义的属性可参考 Android 的帮助文档的 android.R.styleable 类定义中的 Theme 节,图 6-5 展示了部分可用于主题样式的属性。

```
public static final int[] Theme
These are the standard attributes that make up a complete theme.
Includes the following attributes:
```

Attribute	Description
android:absListViewStyle	Default AbsListView style.
android:actionBarDivider	Custom divider drawable to use for elements in the acti
android:actionBarItemBackground	Custom item state list drawable background for action b items.
android:actionBarPopupTheme	Reference to a theme that should be used to inflate pop shown by widgets in the action bar.
android:actionBarSize	Size of the Action Bar, including the contextual bar us present Action Modes.
android:actionBarSplitStyle	Reference to a style for the split Action Bar.
android:actionBarStyle	Reference to a style for the Action Bar
android:actionBarTabBarStyle	

图 6-5　部分可用于主题样式的属性

6.4　使用 Android 平台已定义的样式和主题

Android 平台已经定义了一系列的样式和主题供应用程序使用，在定义的所有样式中，以 Theme 开头的样式是主题样式，其他的不是以 Theme 开头的则是普通样式。Android 完整的样式定义读者可参考 android.R.style 类。想要使用 Android 已定义的样式或主题，需要将样式或主题名中的下画线"_"替换为小数点"."。例如，我们想要在程序的某个 Activity 中使用 Theme_NoTitleBar 主题样式，则需要按如下方式使用：

```
<activity android:theme="@android:style/Theme.NoTitleBar">
```

Android 平台已定义的部分典型的样式和主题如表 6-1 所示。

表 6-1　Android 平台已定义的部分典型的样式和主题

名　称	类　型	描　述
Animation	样式	Android 动画的基础样式
DeviceDefault_ButtonBar	样式	Android 工具条基础样式
Holo_ButtonBar	样式	基于 Holo 风格的工具条样式
MediaButton_(XXX)	样式	媒体播放器样式
TextAppearance	样式	文本显示基础样式
TextAppearance_(XXX)	样式	文本显示变体样式
Theme	主题	Android 基础主题
Theme_Black	主题	Android 黑色主题
Them_Black_NoTitleBar	主题	黑色主题，但不显示 Activity 标题栏
Theme_Dialog	主题	将 Activity 显示为对话框时使用的主题
Theme_Holo	主题	Holo 风格的基础主题
Theme_Holo_Dialog	主题	Holo 风格主题的对话框主题
Theme_Holo_Light	主题	Holo 风格的明亮主题
Theme_Holo_Light_NoActionBar	主题	Holo 风格的明亮主题，但不显示 Activity 的动作栏

续表

名 称	类 型	描 述
Theme_Light	主题	显示明亮背景和黑色字体的主题
Theme_Light_NoTitleBar_FullScreen	主题	全屏、明亮背景、黑色字体、不显示标题栏
Theme_Material	主题	Material 风格的主题
Theme_Material_Light	主题	明亮风格的 Material 主题

6.5 Android 应用程序的主题样式结构分析

介绍完样式与主题的相关知识后，现在回到 Android 程序，来介绍一下 Android 程序中与样式主题相关的内容。

当我们在 Android Studio 中新建一个 Android 工程时，Android 已经设置了默认的主题。打开 AndroidManifest.xml 文件，其内容如下所示：

```xml
<?xml version="1.0" encoding="utf-8"?>
<manifest xmlns:android="http://schemas.android.com/apk/res/android"
    package="com.ttt.ex06styletheme01">

    <application
        android:allowBackup="true"
        android:icon="@mipmap/ic_launcher"
        android:label="@string/app_name"
        android:supportsRtl="true"
        android:theme="@style/AppTheme">  //这里指定了该应用程序的主题
        <activity
            android:name=".MainActivity"
            android:label="@string/app_name"
            android:theme="@style/AppTheme.NoActionBar">  //还有这句
            <intent-filter>
                <action android:name="android.intent.action.MAIN" />

                <category android:name="android.intent.category.LAUNCHER" />
            </intent-filter>
        </activity>
    </application>

</manifest>
```

其中的

```
android:theme="@style/AppTheme" >    //这里指定了该应用程序的主题
```

指定了该应用程序的主题为 AppTheme。我们在 res/values 目录下的 styles.xml 文件中，可以找到对主题样式 AppTheme 的定义。res/values/styles.xml 文件内容如下：

```xml
<resources>

    <!-- Base application theme. -->
    <style name="AppTheme" parent="Theme.AppCompat.Light.NoActionBar">
        <!-- Customize your theme here. -->
        <item name="colorPrimary">@color/colorPrimary</item>
        <item name="colorPrimaryDark">@color/colorPrimaryDark</item>
        <item name="colorAccent">@color/colorAccent</item>
```

```xml
    </style>

    <style name="AppTheme.NoActionBar">
        <item name="windowActionBar">false</item>
        <item name="windowNoTitle">true</item>
    </style>

    <style name="AppTheme.AppBarOverlay" parent="ThemeOverlay.AppCompat.Dark.ActionBar" />

    <style name="AppTheme.PopupOverlay" parent="ThemeOverlay.AppCompat.Light" />

</resources>
```

6.6 同步练习

Android 平台中预定义了很多主题样式,请读者将表 6-1 中的主题样式应用到一个例子程序中,观察一下每个主题样式的外观。

第 7 章

理解和使用 Intent

在基于 HTML 的页面程序中，我们使用"超链接"来实现页面之间的跳转。之前我们也提到过，Android 应用程序界面是由一个或多个 Activity 组成的，一个 Activity 相当于 HTML 的一个页面，那么，当一个 Android 应用程序具有多个相互联系的 Activity 时，它们之间是如何实现跳转的呢？其是通过使用本章将要介绍的 Intent 来实现的。

Intent 的作用不仅是实现 Activity 之间的跳转，它还是 Android 平台的各个部分之间实现信息沟通的桥梁。本章将对 Intent 的使用进行详细的介绍。

7.1 Intent 应用入门案例

我们通过一个简单的例子来说明什么是 Intent 及 Intent 的基本应用。

这个例子程序的目标是：该程序首先显示一个 Activity，在这个 Activity 上使用 TextView 组件显示一张图片的名称及一个按钮，点击这个按钮，将会在一个新的 Activity 中显示这张图片。非常简单，不是吗？但它能很好地帮助读者理解 Intent。

在 Android Studio 中新建一个名为 Ex07Intent01 的 Android 工程，然后修改 res/layout/activity_main.xml，这是该应用程序的主界面，内容如下：

```xml
<?xml version="1.0" encoding="utf-8"?>
<LinearLayout xmlns:android="http://schemas.android.com/apk/res/android"
    android:layout_width="match_parent"
    android:layout_height="match_parent"
    android:orientation="vertical">

    <TextView
        android:layout_width="match_parent"
        android:layout_height="0dp"
        android:layout_weight="1"
        style="@android:style/TextAppearance.Holo.Large"
        android:gravity="center"
        android:text="@string/text_beauty" />

    <View
        android:layout_width="match_parent"
        android:layout_height="10dp"
        android:background="#000"/>

    <Button
```

```xml
    android:id="@+id/id_button"
    android:layout_width="match_parent"
    android:layout_height="0dp"
    android:layout_weight="1"
    android:text="@string/text_button" />

</LinearLayout>
```

这个布局文件很简单，显示一个 TextView 和一个 Button，并在它们之间加上一个 10dp 的间隔。同时，我们对 TextView 还使用了 Android 已定义的文字样式来显示其中的文字，使显示的文字更大。

为了能在一个新的 Activity 中显示一张图片，我们需要为这个 Activity 创建一个布局资源，为此，在 res/layout 目录下，新建一个名为 layout02.xml 的布局文件，其内容如下：

```xml
<?xml version="1.0" encoding="utf-8"?>
<LinearLayout xmlns:android="http://schemas.android.com/apk/res/android"
    android:layout_width="match_parent"
    android:layout_height="match_parent"
    android:orientation="vertical" >

    <ImageView
        android:layout_width="match_parent"
        android:layout_height="match_parent"
        android:src="@drawable/beauty"
        android:scaleType="center"
        android:contentDescription="@string/text_empty" />

</LinearLayout>
```

我们需要将一张图片放置在 res/drawable 资源目录下。由于我们在布局文件中使用了字符串引用，因此，需要修改 res/values/strings.xml 文件，修改后的内容如下：

```xml
<?xml version="1.0" encoding="utf-8"?>
<resources>

    <string name="app_name">Ex07Intent01</string>

    <string name="text_beauty">一张美丽的风景照片</string>
    <string name="text_empty"></string>
    <string name="text_button">点击按钮显示最美风景</string>

</resources>
```

如果此时我们运行这个程序，能够正确显示第一个 Activity 界面，可是当点击按钮时，不会显示第二个 Activity 界面。

为了能显示第二个 Activity 界面，我们需要创建一个新的 Activity，使之显示 layout2.xml 布局的界面。为此，在 Ex07Intent01 工程的 src 目录的 com.ttt.ex07intent01 包下，新建一个名为 Activity02 的 Java 类文件，Activity02.java 文件内容如下：

```java
package com.ttt.ex07intent01;

import android.app.Activity;
import android.os.Bundle;
```

```java
public class Activity02 extends Activity {

    @Override
    protected void onCreate(Bundle savedInstanceState) {
        super.onCreate(savedInstanceState);
        setContentView(R.layout.layout02);
    }
}
```

这个 Activity 只显示 layout2.xml 布局文件的界面。

现在有两个 Activity：MainActivity 和 Activity02，它们分别显示 activity_main.xml 和 layout02 布局。可是如何把它们关联起来呢？也就是如何在点击第一个界面上的按钮时，启动第二个 Activity 运行来显示 layout2 布局呢？为此，需要修改 MainActivity.java，使之监听对按钮的点击事件，并对点击事件进行处理，修改后的 MainActivity.java 文件的内容如下：

```java
package com.ttt.ex07intent01;

import android.content.Intent;
import android.os.Bundle;
import android.support.v7.app.AppCompatActivity;
import android.view.View;
import android.widget.Button;

public class MainActivity extends AppCompatActivity implements View.OnClickListener {

    @Override
    protected void onCreate(Bundle savedInstanceState) {
        super.onCreate(savedInstanceState);
        setContentView(R.layout.activity_main);

        Button btn = (Button)this.findViewById(R.id.id_button);
        btn.setOnClickListener(this);
    }

    @Override
    public void onClick(View v) {
        Intent i = new Intent(this, Activity02.class);
        this.startActivity(i);
    }

}
```

在对按钮的点击处理代码中，我们首先创建了一个 Intent 对象实例，然后通过 startActivity(i) 启动指定的 Activity 来运行程序。在运行程序之前，还需要声明 Activity02 中的 Activity，因为，Android 规定，所有的 Activity 都必须在工程 AndroidManifest.xml 文件中进行登记。为此，修改 AndroidManifest.xml 文件，内容如下：

```xml
<?xml version="1.0" encoding="utf-8"?>
<manifest xmlns:android="http://schemas.android.com/apk/res/android"
    package="com.ttt.ex07intent01">

    <application
        android:allowBackup="true"
```

```xml
        android:icon="@mipmap/ic_launcher"
        android:label="@string/app_name"
        android:supportsRtl="true"
        android:theme="@style/AppTheme">
        <activity
            android:name=".MainActivity"
            android:label="@string/app_name"
            android:theme="@style/AppTheme.NoActionBar">
            <intent-filter>
                <action android:name="android.intent.action.MAIN" />

                <category android:name="android.intent.category.LAUNCHER" />
            </intent-filter>
        </activity>

        <activity
            android:name=".Activity02"
            android:label="@string/app_name"
            android:theme="@style/AppTheme.NoActionBar">
        </activity>
    </application>
</manifest>
```

在 AndroidManifest.xml 文件中，我们在<application>标签下，通过<activity>标签来登记新的 Activity02。现在运行该程序，将显示如图 7-1 所示的界面。

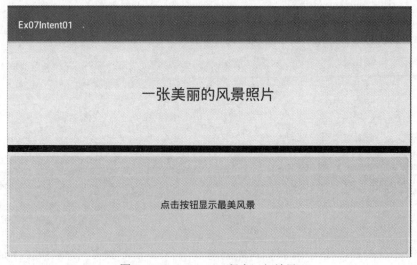

图 7-1　Ex07Intent01 程序运行效果

点击按钮，将显示一张美丽的风景照片，如图 7-2 所示。

第 7 章 理解和使用 Intent

图 7-2 点击按钮后的界面

这正是我们期望的结果。

7.2 同步练习一

编写一套与 7.1 节中的例子相似的、能通过 Intent 打开新的 Activity 的程序。例如，读者可以通过点击一张图片来显示一个文本框，该文本框用来介绍该图片的一些基本情况。

7.3 细说 Intent

Android 的 Intent 对象是联系各个 Activity 的关键对象。Intent，翻译成中文就是"意图"，我们可以这样来理解 Intent：通过 Intent 对象，来告诉 Android 要做什么。例如，在 7.1 节的例子中，我们构建一个 Intent：

```
Intent i = new Intent(this, Activity02.class);
this.startActivity(i);
```

意思就是说，我们希望 Android 平台替我们打开 Activity02 界面。按照这种方式创建的 Intent 称为显式 Intent：在 Intent 中，我们明确地告诉 Android 系统要启动的 Activity。还有一种 Intent 称为隐式 Intent：在 Intent 中指定一些条件，由 Android 系统根据这些条件来启动最能满足条件的 Activity。

为了便于读者理解隐式 Intent，我们修改 7.1 节中的例子。首先，修改 AndroidManifest.xml 为如下内容：

```xml
<?xml version="1.0" encoding="utf-8"?>
<manifest xmlns:android="http://schemas.android.com/apk/res/android"
   package="com.ttt.ex07intent01">

   <application
      android:allowBackup="true"
      android:icon="@mipmap/ic_launcher"
      android:label="@string/app_name"
      android:supportsRtl="true"
```

```xml
        android:theme="@style/AppTheme">
    <activity
        android:name=".MainActivity"
        android:label="@string/app_name"
        android:theme="@style/AppTheme.NoActionBar">
        <intent-filter>
            <action android:name="android.intent.action.MAIN" />

            <category android:name="android.intent.category.LAUNCHER" />
        </intent-filter>
    </activity>

    <activity
        android:name=".Activity02">
        <intent-filter>
            <action android:name="com.ttt.ex07intent01.A1" />
            <category android:name="android.intent.category.DEFAULT" />
        </intent-filter>
    </activity>

</application>

</manifest>
```

读者注意下面的代码：

```xml
<intent-filter>
    <action android:name="com.ttt.ex07intent01.A1" />
    <category android:name="android.intent.category.DEFAULT" />
</intent-filter>
```

这段代码为 Activity02 中的 Activity 定义了一个 Intent 过滤器，它表示：在创建 Intent 对象时，只要指定 Intent 的 action 的名字为 "com.ttt.ex07intent01.A1"，并且指明 Intent 的 category 为 "android.intent.category.DEFAULT"，即可打开 Activity02 界面。同时，Android 也允许为一个 Activity 定义多个 Intent 过滤器。为了能够采用隐式的方式打开 Activity02，修改 MainActivity.java 为如下内容：

```java
package com.ttt.ex07intent01;

import android.content.Intent;
import android.os.Bundle;
import android.support.v7.app.AppCompatActivity;
import android.view.View;
import android.widget.Button;

public class MainActivity extends AppCompatActivity implements View.OnClickListener {

    @Override
    protected void onCreate(Bundle savedInstanceState) {
        super.onCreate(savedInstanceState);
        setContentView(R.layout.activity_main);

        Button btn = (Button)this.findViewById(R.id.id_button);
        btn.setOnClickListener(this);
```

```
    }

    @Override
    public void onClick(View v) {
        Intent i = new Intent("com.ttt.ex07intent01.A1");
        i.addCategory(Intent.CATEGORY_DEFAULT);
        this.startActivity(i);
    }
}
```

注意创建 Intent 和启动 Activity02 的代码：

```
Intent i = new Intent("com.ttt.ex07intent01.A1");
i.addCategory(Intent.CATEGORY_DEFAULT);
this.startActivity(i);
```

在这段代码中，在创建 Intent 时指明了要打开的 Activity 支持的 action 的名称，同时通过 addCategory()方法指定了当前动作（Action）被执行的环境，因此，当执行 this.startActivity(i) 语句时，Android 平台将打开能满足这些条件的 Activity，此处就是 Activity02。这里有一个问题：当在 Intent 中有多个 Activity 能满足指明的条件时，Android 平台会怎样处理呢？答案是：Android 平台将弹出一个对话框，让我们选择满足条件的 Activity 来运行。

运行这个程序，会得到与图 7-1 和图 7-2 一致的结果。

从这个例子我们可以看出，AndroidManifest.xml 文件对 Activity 的配置，特别是<activity>标签的<intent-filter>子标签所指定的属性与创建 Intent 对象时所指定的各种属性是密切相关的：它们必须匹配才能打开我们需要的 Activity。它们之间的关系如图 7-3 所示。

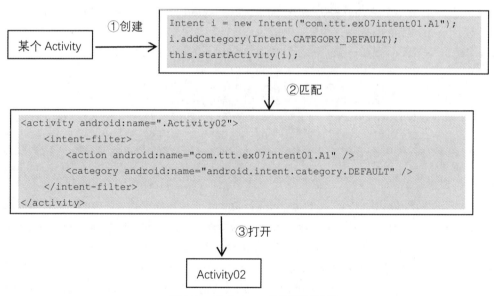

图 7-3　隐式 Intent 的配置和使用

图 7-3 表示：某个 Activity 首先在 AndroidManifest.xml 文件中配置自己，包括配置自己的<intent-filter>，然后，另一个 Activity 创建一个 Intent 对象，在创建的 Intent 对象中所指定的 action 及 category 正是之前的那个 Activity 在其<intent-filter>中所设定的，所以，当调用 startActivity(i)时，则会打开该 Intent 所指定的那个 Activity。

那么，在创建 Intent 时，可以指明哪些条件，或者说，在为 Activity 定义<intent-filter>标签时，可以定义哪些条件呢？可以指明的条件包括 action、data 和 category。下面笔者将分别进行讲解。

7.3.1 Intent 的 action

我们可以为创建的 Activity 定义一个用于打开 Activity 的 action，action 是一个字符串常量，我们可以任意定义，但是，Android 建议的做法是，采用"Java 包名+特定串"的形式来命名 action。例如，我们为 Activity02 定义的 action 为：com.ttt.ex07intent01.A1。其中，com.ttt.ex07intent01 是包名，而 A1 是特定的名称。这里需要强调的是，Activity 与 action 不一定是 1 对 1 的关系，也就是说，可能多个 Activity 对应同名的 action，这在 Android 中是允许的，当出现这种情况时，Android 平台将弹出一个对话框，让我们选择满足条件的 Activity 来运行。

在 Intent 类中已经预定义了一些常用的 action，包括 Intent.ACTION_MAIN、Intent.ACTION_VIEW、Intent.ACTION_EDIT、Intent.ACTION_PICK、Intent.ACTION_DIAL、Intent.ACTION_CALL、Intent.ACTION_DELETE、Intent.ACTION_INSERT、Intent.ACTION_SEARCH 等。其中的每个 action 都有确定的含义，例如，Intent.ACTION_MAIN 表示包含这个 action 的 Activity 是 Android 应用程序的入口 Activity，也就是，当运行某个 Android 应用程序时，首先打开包含这个 action 的 Activity；Intent.ACTION_PICK 表示从某个列表中选择一条信息。

这些 action 可以用来打开系统已经包含的一些 Activity，如电话拨号 Activity。我们也可以在程序中直接使用这些 action 作为 Activity 的 action。指明一个 Intent 的 action，有以下两种方式。

（1）通过 Intent 类的构造函数 new Intent(String action)及 new Intent(String action, URI uri)。

（2）通过 Intent 类的 setAction(String action)函数。

同时，想要指明一个 Activity 能被哪一个或哪些 action 打开，需要在 AndroidManifest.xml 文件中，使用<intent-filter>标签来说明，内容如下：

```xml
<activity class=".NotesList" android:label="@string/title_notes_list">
    <intent-filter>
        <action android:name="android.intent.action.VIEW" />
        <action android:name="android.intent.action.EDIT" />
        <action android:name="android.intent.action.PICK" />
        <category android:name="android.intent.category.DEFAULT" />
        <data android:mimeType="vnd.android.cursor.dir/vnd.google.note" />
    </intent-filter>
</activity>
```

这个例子说明，NotesList 中的 Activity 只有在 Intent 的 action 中设定为 android.intent.action.VIEW、android.intent.action.EDIT 和 android.intent.action.PICK 中的 1 个时，action 才能被打开，注意，当我们在 AndroidManifest.xml 文件中设置 action 的 android:name 属性时，需要使用常量名称所对应的字符串值，而不是常量名称。同时，想要打开这个 Activity，还需要在 Intent 的 category 中设置 android.intent.category.DEFAULT 及在 Intent 的 data 中设置 mimeType 为 vnd.android.cursor.dir/vnd.google.note。接下来我们将介绍 Intent 的这两个数据项。

7.3.2 Intent 的 data

在通过隐式 Intent 打开 Activity 时，除了指明 Activity 的 action，我们经常还指明 Activity 所支持的 data。就像在 HTML 中那样，除了指明 GET、PUT、POST 动作，还经常需要指明页面的 URI 地址。在 Android 中，通过 Intent 的 data 指明要操作的数据，Android 中的 data 也是通过 URI 来指明的。

URI 的标准形式为：scheme://host:port/path、scheme://host:port/pathPattern 或 scheme://host:port/pathPrefix。

例如，在如下的 URI 中：content://com.example.project:200/folder/subfolder/etc。scheme 为"content"，host 为"com.example.project"，port 为"200"，path 为"folder/subfolder/etc"。

想要指明某个 Activity 的 data 属性，我们需要在 AndroidManifest.xml 文件中，在 Activity 配置的<intent-filter>子标签中通过<data>标签指定，指定 intent-filter 的 data 属性的一般形式如下：

```xml
<data android:host="string"
    android:mimeType="string"
    android:path="string"
    android:pathPattern="string"
    android:pathPrefix="string"
    android:port="string"
    android:scheme="string" />
```

通过<activity>标签的子标签<intent-filter>的<data>子标签，我们可以指定某个 Activity 的 data 属性，其中包括 host、mimeType、path、pathPattern、pathPrefix、port 和 scheme。

为了通过 Intent 打开设定了 data 属性的某个 Activity，需要在创建 Intent 对象时，明确设置 data 属性。在 Intent 类中设置 data 属性的方法如下。

（1）setData(Uri data)。

设置 Intent 的 data 属性，注意这个方法将自动清除之前设置的 mimeType。

（2）setDataAndType(Uri data, String mimeType)。

同时设置 Intent 的 data 和 mimeType。

（3）setType(String mimeType)。

设置 Intent 的 mimeType，这个方法将自动清除之前设置的 data 属性。

下面举一个例子来说明如何在<intent-filter>标签中设置 data 属性，并使用隐式 Intent 来打开设置了 data 属性值的 Activity。

首先在 AndroidManifest.xml 文件中配置某个 Activity，内容如下：

```xml
<activity class=".NotesList" android:label="@string/title_notes_list">
    <intent-filter>
        <action android:name="android.intent.action.MAIN" />
        <category android:name="android.intent.category.LAUNCHER" />
    </intent-filter>
    <intent-filter>
        <action android:name="android.intent.action.VIEW" />
        <action android:name="android.intent.action.EDIT" />
        <action android:name="android.intent.action.PICK" />
        <category android:name="android.intent.category.DEFAULT" />
        <data android:scheme="content" />
        <data android:mimeType="vnd.android.cursor.dir/vnd.google.note" />
    </intent-filter>
</activity>
```

在第一个<intent-filter>标签中，我们配置了其 action 为 android.intent.action.MAIN，这说明，该 Activity 是这个 Activity 所属的 Android 程序的主入口 Activity，同时，在第一个<intent-filter>

标签中，我们也配置了该 Activity 的 category 为 android.intent.category.LAUNCHER，这说明，该 Activity 将出现在 Android 的 HOME 界面的应用程序列表中。再来看第二个<intent-filter>标签，我们配置了该 Activity 的 action 为 android.intent.action.VIEW、android.intent.action.EDIT 和 android.intent.action.PICK，category 为 android.intent.category.DEFAULT，并且，配置其 data 的 scheme 为 content，mimeType 为 vnd.android.cursor.dir/vnd.google.note。我们可以通过创建如下的 Intent 对象来通过隐式 Intent 打开这个 Activity：

```
Uri uri = Uri.parse("content://com.example.project:200/folder/subfolder/etc");
Intent i = new Intent("android.intent.action.EDIT");
i.setDataAndType(uri, "vnd.android.cursor.dir/vnd.google.note");
this.startActivity(i);
```

这有一个问题：我们在配置中已经设定了 Activity 的 category，但是在创建 Intent 对象时，并没有设置 category 属性，这能打开配置了 category 为 DEFAULT 的 Activity 吗？答案是：能。为什么呢？因为，Android 在创建 Intent 对象时，会自动设定 Intent 的 category 为 DEFAULT。那么什么是 category，我们将在 7.3.3 节介绍。

7.3.3 Intent 的 category

category，顾名思义，就是类别的意思，它是 Android 对 Activity 进行分类的一种手段。例如，如果我们设定某个 Activity 的 category 为 LAUNCHER，那么，Android 会在启动时将它显示在 HOME 屏幕上以便于运行它；如果我们设定某个 Activity 的 category 为 HOME，那么，Android 在启动时会把它作为 Android 系统的 HOME 屏幕。

Android 已经在 Intent 类中定义了一些 category 常量，常用的 category 包括 Intent.CATEGORY_DEFAULT、Intent.CATEGORY_LAUNCHER、Intent.CATEGORY_INFO、Intent.CATEGORY_HOME、Intent.CATEGORY_PREFERENCE、Intent.CATEGORY_CAR_DOCK、Intent.CATEGORY_CAR_MODE、Intent.CATEGORY_APP_MARKET 等。

下面举一个例子，这个例子可以获得所有出现在 HOME 屏幕上的 Activity 列表，内容如下：

```
Intent mainIntent = new Intent(Intent.ACTION_MAIN, null);
mainIntent.addCategory(Intent.CATEGORY_LAUNCHER);
PackageManager pm = getPackageManager();
List<ResolveInfo> list = pm.queryIntentActivities(mainIntent, 0);
for(ResolveInfo ri: list)
{
   Log.d("test",ri.toString());
   String packagename = ri.activityInfo.packageName;
   String classname = ri.activityInfo.name;
   Log.d("test", packagename + ":" + classname);
   if (classname.equals("com.ttt.ex07intent01.Activity02"))
   {
      Intent ni = new Intent();
      ni.setClassName(packagename,classname);
      activity.startActivity(ni);
   }
}
```

这个例子程序的编写思路是：首先创建一个 action 为 Intent.ACTION_MAIN 和 category 为 Intent.CATEGORY_LAUNCHER 的 Intent 对象，然后通过 PackageManager 查询所有匹配该

Intent 的 Activity，并显示匹配 Activity 的类名，如果一个 Activity 的类名等于指定的类名：com.ttt.ex07intent01.Activity02，则打开该 Activity 运行。

再举一个例子。通过下面的程序片段，可以在程序中直接回到 HOME 界面：

```
Intent mainIntent = new Intent(Intent.ACTION_MAIN, null);
mainIntent.addCategory(Intent.CATEGORY_HOME);
startActivity(mainIntent);
```

7.3.4 Intent 的 extra

在通过 Intent 启动某个 Activity 运行时，有时可能需要传递一些附加的数据到被启动的 Activity 中，这可以使用 extra。extra 只作为传递给目标 Activity 的附加数据，不作为挑选 Activity 的匹配依据。

extra 是以 "key/value" 形式表示的数据，其中的 key 是 String 类型的 "键"，value 可以是 Java 基本数据类型也可以是实现了 android.os.Parcelable 接口的对象数据类型。Intent 类提供了写入及读取基本数据类型数据的方法。例如，我们希望传递一个整数和一个字符串到目标 Activity，则可以在创建的 Intent 对象上执行如下的代码：

```
intent.putExtra("productName", "iPhone");
intent.putExtra("ProductAmount", 100);
```

在目标 Activity 中，通过获得打开该 Activity 的 Intent 对象来从中获得传递过来的数据，内容如下：

```
String pn = intent.getStringExtra("productName");
int pa = intent.getIntExtra("ProductAmount");
```

7.4 Intent 解析

从上面的介绍可以看出，Intent 对象与<intent-filter>标签是密切相关的：在通过 Intent 对象来启动某个 Activity 时，必须将 Intent 对象中所设定的属性（action、data 和 category）与<intent-filter>标签中所配置的属性进行匹配，从而打开能够匹配的 Activity。我们把这种匹配操作称为 Intent 解析。Android 按如下步骤来解析 Intent。

（1）在显式匹配中，在 Intent 对象中明确指定要打开的 Activity 的类的名称。包括 new Intent(Context packageContext, Class<?> cls)、setComponent(ComponentName)、setClass(Context, Class)。

（2）在隐式匹配中，分为如下几种情况。

- 如果在 Intent 中设定了 action，且 action 出现在<intent-filter>标签中或者在<intent-filter>标签中未设定任何 action，则匹配；如果在 Intent 中未设定 action，则匹配所有的 Activity。
- 如果在 Intent 中设定了 data，且 data 属性出现在<intent-filter>标签的 data 列表中，则匹配；如果在 Intent 中未设定 data，则只能匹配未设定任何 data 属性的 Activity。
- 如果在 Intent 中设定了 category，且 category 必须在<intent-filter>标签的 category 列表中，则匹配；如果在 Intent 中未设定 category，则只能匹配未设定任何 category 的 Activity。

7.5 获得 Activity 返回的结果

在前面的例子中通过 Intent 打开某个 Activity 时，并不需要被打开的目标 Activity 返回结果，而有时我们确实需要获得被打开的 Activity 返回的结果，该怎么做呢？

想要获得被打开的 Activity 返回的结果，我们不是使用 startActivity()方法，而是应该使用 startActivityForResult()方法。通过 startActivityForResult()启动 Activity 时，当被启动的 Activity（称为目标 Activity）执行完毕时，Android 平台将调用源 Activity 的 onActivityResult()方法。

startActivityForResult()方法的原型如下：

```
public void startActivityForResult(Intent intent, int requestCode)
```

其中的 requestCode 是源 Activity 传递的一个识别参数，当目标 Activity 执行完毕，Android 平台调用 onActivityResult()方法，onActivityResult()方法的原型如下：

```
protected void onActivityResult(int requestCode, int resultCode, Intent data)
```

其中的 requestCode 就是在调用 startActivityForResult()方法时所传进去的 requestCode 参数，而 resultCode 则是执行目标 Activity 的结果，其值可以为：Activity.RESULT_OK、Activity.RESULT_CANCELED 或其他自定义的值。

当然，要获得目标 Activity 的返回结果的前提是：目标 Activity 首先要返回结果。如果目标 Activity 根本就没有返回结果，源 Activity 也就无从获得返回结果了。目标 Activity 采用 setResult()方法来设置返回数据。setResult()方法的原型如下：

```
public final void setResult(int resultCode)
public final void setResult(int resultCode, Intent data)
```

也就是说，目标 Activity 也是通过 Intent 对象来返回结果数据给源 Activity 的。

下面，我们举一个例子来说明如何启动一个 Activity，并接收从目标 Activity 返回的结果。这个程序首先显示如图 7-4 所示的界面。

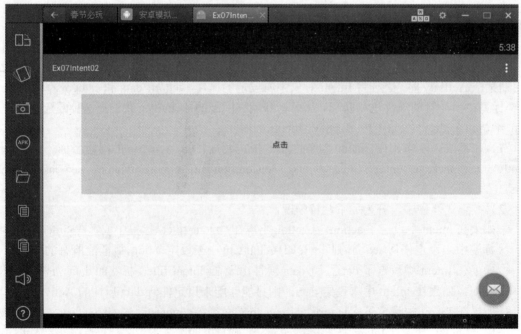

图 7-4　程序运行首界面

点击按钮，界面将显示两张图片，如图 7-5 所示。

此时，点击任何一张图片，将关闭该 Activity，并在首界面上显示所点击图片的编号，如图 7-6 所示。

第 7 章 理解和使用 Intent

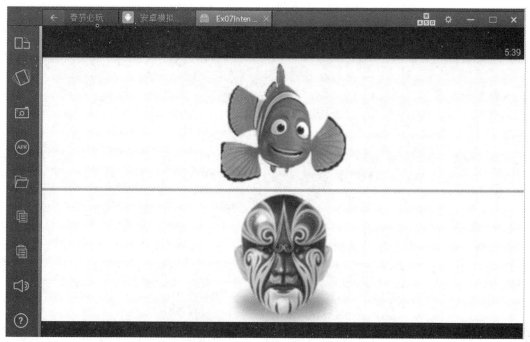

图 7-5 点击首界面的按钮后显示的界面

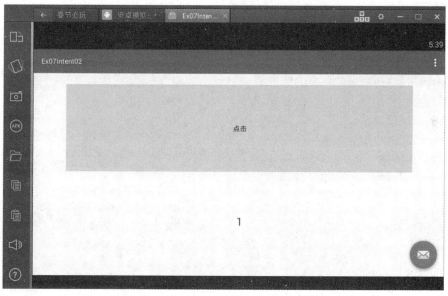

图 7-6 在主界面上显示所点击图片的编号

现在介绍该程序的代码。首先在 Android Studio 中创建一个名为 Ex07Intent02 的 Android 工程。修改 res/layout/activity_main.xml 主布局文件，内容如下：

```xml
<?xml version="1.0" encoding="utf-8"?>
<LinearLayout xmlns:android="http://schemas.android.com/apk/res/android"
    android:layout_width="match_parent"
    android:layout_height="match_parent"
    android:orientation="vertical">

    <Button
        android:id="@+id/id_button"
```

```
        android:layout_width="match_parent"
        android:layout_height="0dp"
        android:layout_weight="1"
        android:text="@string/text_button" />

    <TextView
        android:id="@+id/id_textview"
        android:layout_width="match_parent"
        android:layout_height="0dp"
        android:layout_weight="1"
        style="@android:style/TextAppearance.Holo.Large"
        android:gravity="center" />

</LinearLayout>
```

然后在 res/layout 目录下，新建一个名为 activity_second.xml 的布局文件，内容如下：

```
<?xml version="1.0" encoding="utf-8"?>
<LinearLayout xmlns:android="http://schemas.android.com/apk/res/android"
    android:layout_width="match_parent"
    android:layout_height="match_parent"
    android:orientation="vertical" >

    <ImageView
        android:id="@+id/id_iv01"
        android:layout_width="match_parent"
        android:layout_height="0dp"
        android:layout_weight="1"
        android:scaleType="fitCenter"
        android:src="@drawable/png0019"
        android:contentDescription="@string/text_empty"/>

    <View
        android:layout_width="match_parent"
        android:layout_height="4dp"
        android:background="#aaa" />

    <ImageView
        android:id="@+id/id_iv02"
        android:layout_width="match_parent"
        android:layout_height="0dp"
        android:layout_weight="1"
        android:scaleType="fitCenter"
        android:src="@drawable/png0021"
        android:contentDescription="@string/text_empty"/>

</LinearLayout>
```

同时在 res/drawable 目录下，放置两个需要在该布局文件中用到的图片资源文件：png0019.png 文件和 png0021.png 文件。接下来修改 res/values/strings.xml 文件，在其中添加布局文件中用到的字符串引用，修改后的内容如下：

```
<?xml version="1.0" encoding="utf-8"?>
<resources>
```

```xml
    <string name="app_name">Ex07Intent02</string>

    <string name="text_button">点击</string>
    <string name="text_empty"></string>

</resources>
```

现在修改 MainActivity.java 文件，使其显示主界面 activity_main.xml 布局，并监听对按钮的点击事件，修改后的内容如下：

```java
package com.ttt.ex07intent02;

import android.app.Activity;
import android.content.Intent;
import android.os.Bundle;
import android.support.v7.app.AppCompatActivity;
import android.view.View;
import android.widget.Button;
import android.widget.TextView;

public class MainActivity extends AppCompatActivity implements View.OnClickListener {
    public static int RC01 = 1000;

    private Button btn;
    private TextView tv;

    @Override
    protected void onCreate(Bundle savedInstanceState) {
        super.onCreate(savedInstanceState);
        setContentView(R.layout.activity_main);

        btn = (Button)this.findViewById(R.id.id_button);
        btn.setOnClickListener(this);

        tv = (TextView)this.findViewById(R.id.id_textview);

    }

    @Override
    public void onClick(View v) {
        Intent intent = new Intent(this, SecondActivity.class);
        this.startActivityForResult(intent, MainActivity.RC01);
    }

    @Override
    protected void onActivityResult(int requestCode, int resultCode, Intent data) {
        if ((requestCode == MainActivity.RC01) && (resultCode == Activity.RESULT_OK)) {
            int which = data.getIntExtra("result", -1);
            tv.setText(which + "");
        }
    }

}
```

在 onClick()方法中，我们通过 startActivityForResult()方法打开 SecondActivity，并传递一个请求码 MainActivity.RC01，代码如下：

```
@Override
public void onClick(View v) {
    Intent intent = new Intent(this, SecondActivity.class);
    this.startActivityForResult(intent, MainActivity.RC01);
}
```

当 SecondActivity 返回时，Android 会调用 MainActivity 的 onActivityResult()方法，在这个方法中，我们获得 SecondActivity 的返回值，并将该值显示在 MainActivity 的 TextView 组件中，内容如下：

```
@Override
protected void onActivityResult(int requestCode, int resultCode, Intent data) {
    if ((requestCode == MainActivity.RC01) && (resultCode == Activity.RESULT_OK)) {
        int which = data.getIntExtra("result", -1);
        tv.setText(which + "");
    }
}
```

最后，我们在 src 目录的 com.ttt.ex07intent02 包下，新建一个名为 SecondActivity 的 Java 类文件，在其中显示 activity_second.xml 布局，并加入图片的点击事件处理方法。SecondActivity.java 文件的内容如下：

```
package com.ttt.ex07intent02;

import android.app.Activity;
import android.content.Intent;
import android.os.Bundle;
import android.view.View;
import android.view.View.OnClickListener;
import android.widget.ImageView;

public class SecondActivity extends Activity implements OnClickListener {
    private ImageView iv01,iv02;

    @Override
    protected void onCreate(Bundle savedInstanceState) {
        super.onCreate(savedInstanceState);
        setContentView(R.layout.activity_second);

        iv01 = (ImageView)this.findViewById(R.id.id_iv01);
        iv01.setOnClickListener(this);
        iv02 = (ImageView)this.findViewById(R.id.id_iv02);
        iv02.setOnClickListener(this);
    }

    @Override
    public void onClick(View v) {
        int id = v.getId();
        if (id == R.id.id_iv01) {
            Intent intent = new Intent();
            intent.putExtra("result", 1);
            this.setResult(Activity.RESULT_OK, intent);
```

```
        } else {
            Intent intent = new Intent();
            intent.putExtra("result", 2);
            this.setResult(Activity.RESULT_OK, intent);
        }

        this.finish();
    }
}
```

在 SecondActivity 的 onClick()方法中,首先判断被点击的图片,然后创建一个 Intent 对象来给源 Activity 传递返回数据,最后调用 SecondActivity 的 finish()方法来关闭 SecondActivity。

同时,我们还需要在 AndroidManifest.xml 文件中配置 SecondActivity,为此,修改 AndroidManifest.xml 文件,内容如下:

```xml
<?xml version="1.0" encoding="utf-8"?>
<manifest xmlns:android="http://schemas.android.com/apk/res/android"
    package="com.ttt.ex07intent02">

    <application
        android:allowBackup="true"
        android:icon="@mipmap/ic_launcher"
        android:label="@string/app_name"
        android:supportsRtl="true"
        android:theme="@style/AppTheme">
        <activity
            android:name=".MainActivity"
            android:label="@string/app_name"
            android:theme="@style/AppTheme.NoActionBar">
            <intent-filter>
                <action android:name="android.intent.action.MAIN" />

                <category android:name="android.intent.category.LAUNCHER" />
            </intent-filter>
        </activity>

        <activity android:name=".SecondActivity">
        </activity>

    </application>

</manifest>
```

完成后运行该程序,即可显示如图 7-4 所示的结果。

7.6 Intent 的综合应用举例

7.6.1 运行效果

本例中将首先显示一个功能列表,当点击某个功能按钮时再执行特定的功能。运行效果如图 7-7 所示。

图 7-7 程序运行首界面

点击"从百度搜索内容"图片按钮,将启动浏览器,如图 7-8 所示。

图 7-8 启动浏览器

点击"拨打电话"图片按钮,将显示拨打电话的界面,如图 7-9 所示。

图 7-9 拨打电话界面

7.6.2 程序代码

首先在 Android Studio 中建立名为 Ex07Intent03 的工程,并把需要用到的图片放置在 res/drawable 目录下。修改 res/layout/activity_main.xml 布局文件,内容如下:

```xml
<?xml version="1.0" encoding="utf-8"?>
<ScrollView xmlns:android="http://schemas.android.com/apk/res/android"
    android:layout_width="match_parent"
    android:layout_height="match_parent">

    <TableLayout
        android:layout_width="match_parent"
        android:layout_height="wrap_content"
        android:stretchColumns="1" >

        <TableRow>
            <Button
                android:id="@+id/button1"
                android:layout_width="wrap_content"
                android:layout_height="wrap_content"
                android:background="@drawable/png0017" />

            <TextView
                android:layout_gravity="center_vertical"
                android:paddingLeft="4dp"
                android:paddingRight="4dp"
                android:text="@string/text_baidu" />
        </TableRow>

        <TableRow>
            <Button
                android:id="@+id/button2"
                android:layout_width="wrap_content"
                android:layout_height="wrap_content"
                android:background="@drawable/png0993" />

            <TextView
                android:layout_gravity="center_vertical"
                android:paddingLeft="4dp"
                android:paddingRight="4dp"
                android:text="@string/text_sina_web" />
        </TableRow>

<TableRow>
            <Button
                android:id="@+id/button3"
                android:layout_width="wrap_content"
                android:layout_height="wrap_content"
                android:background="@drawable/png0647" />

            <TextView
                android:layout_gravity="center_vertical"
                android:paddingLeft="4dp"
```

```xml
            android:paddingRight="4dp"
            android:text="@string/text_dial" />
</TableRow>

<TableRow>
    <Button
        android:id="@+id/button4"
        android:layout_width="wrap_content"
        android:layout_height="wrap_content"
        android:background="@drawable/png0081" />

    <TextView
        android:layout_gravity="center_vertical"
        android:paddingLeft="4dp"
        android:paddingRight="4dp"
        android:text="@string/text_map" />
</TableRow>

<TableRow>
    <Button
        android:id="@+id/button5"
        android:layout_width="wrap_content"
        android:layout_height="wrap_content"
        android:background="@drawable/png0006" />

    <TextView
        android:layout_gravity="center_vertical"
        android:paddingLeft="4dp"
        android:paddingRight="4dp"
        android:text="@string/text_short_msg" />
</TableRow>

<TableRow>
    <Button
        android:id="@+id/button6"
        android:layout_width="wrap_content"
        android:layout_height="wrap_content"
        android:background="@drawable/png0064" />

    <TextView
        android:layout_gravity="center_vertical"
        android:paddingLeft="4dp"
        android:paddingRight="4dp"
        android:text="@string/text_email" />
</TableRow>

<TableRow>
    <Button
        android:id="@+id/button7"
        android:layout_width="wrap_content"
        android:layout_height="wrap_content"
        android:background="@drawable/png0643" />
```

```xml
        <TextView
            android:layout_gravity="center_vertical"
            android:paddingLeft="4dp"
            android:paddingRight="4dp"
            android:text="@string/text_multimedia" />
    </TableRow>
  </TableLayout>

</ScrollView>
```

在这里，我们采用了 TableLayout 的布局方式来显示每行的按钮和文字，由于表格的行数可能会超出物理屏幕的高度，因此，在 TableLayout 的外面我们设置了一个 ScrollView 来实现垂直滚动。一般来说，在比较规整的布局中，可以采用 TableLayout 与 TableRow 相结合的方式来实现行列的规整布局：在 TableLayout 中包含 TableRow，其中，TableRow 表示表格中的一行。TableLayout 的几个 XML 配置参数及其含义如下。

（1）android:stretchColumns。

当显示一行数据时，如果物理屏幕宽度有富余的空间，则将指定的列的宽度自动延展以使用富余的显示空间。

（2）android:shrinkColumns。

当显示一行数据时，如果物理屏幕宽度不足以显示所有列的数据，则自动压缩指定列的宽度以使其他列有足够的显示空间。

（3）android:collapseColumns。

将指定的列不显示出来。

以上 3 个属性都支持以逗号","分隔多个列号，列号从 0 开始。

当然，我们还需要修改 res/values/strings.xml 文件，在其中添加布局文件中用到的字符串引用，修改后的 res/values/strings.xml 文件内容如下：

```xml
<?xml version="1.0" encoding="utf-8"?>
<resources>

    <string name="app_name">Ex07Intent03</string>
    <string name="action_settings">Settings</string>

    <string name="text_baidu">从百度搜索内容</string>
    <string name="text_sina_web">浏览网页\nwww.sina.com.cn</string>
    <string name="text_dial">拨打电话</string>
    <string name="text_map">显示地图</string>
    <string name="text_short_msg">发短信</string>
    <string name="text_email">发送 Email</string>
    <string name="text_multimedia">播放多媒体</string>

</resources>
```

现在来修改 MainActivity.java，使之显示主界面，监听对按钮的点击事件，并根据所点击的按钮，创建满足功能条件的 Intent 对象，来启动相应的 Activity 运行。这个例子也说明，采用 Intent 启动的 Activity 并不一定是自己的 Activity，也可以是第三方程序的 Activity。MainActivity.java 文件内容如下：

```java
package com.ttt.ex07intent03;
```

```java
import android.app.SearchManager;
import android.content.Intent;
import android.net.Uri;
import android.os.Bundle;
import android.provider.MediaStore;
import android.support.v7.app.AppCompatActivity;
import android.view.View;
import android.widget.Button;

public class MainActivity extends AppCompatActivity  implements View.OnClickListener {
    Button btn1,btn2,btn3,btn4,btn5,btn6,btn7;

    @Override
    protected void onCreate(Bundle savedInstanceState) {
        super.onCreate(savedInstanceState);
        setContentView(R.layout.activity_main);

        btn1 = (Button)this.findViewById(R.id.button1);
        btn1.setOnClickListener(this);

        btn2 = (Button)this.findViewById(R.id.button2);
        btn2.setOnClickListener(this);

        btn3 = (Button)this.findViewById(R.id.button3);
        btn3.setOnClickListener(this);

        btn4 = (Button)this.findViewById(R.id.button4);
        btn4.setOnClickListener(this);

        btn5 = (Button)this.findViewById(R.id.button5);
        btn5.setOnClickListener(this);

        btn6 = (Button)this.findViewById(R.id.button6);
        btn6.setOnClickListener(this);

        btn7 = (Button)this.findViewById(R.id.button7);
        btn7.setOnClickListener(this);
    }


    @Override
    public void onClick(View v) {
        Uri uri;
        Intent it;
        switch(v.getId()) {
            case R.id.button1:
                Intent intent = new Intent();
                intent.setAction(Intent.ACTION_WEB_SEARCH);
                intent.putExtra(SearchManager.QUERY,"Andoid应用编程");
                startActivity(intent);
                break;
            case R.id.button2:
                uri = Uri.parse("http://www.sina.com.cn");
```

```
                it = new Intent(Intent.ACTION_VIEW,uri);
                startActivity(it);
                break;
            case R.id.button3:
                uri = Uri.parse("tel:5556");
                it = new Intent(Intent.ACTION_DIAL, uri);
                startActivity(it);
                break;
            case R.id.button4:
                uri = Uri.parse("geo:38.899533,-77.036476");
                it = new Intent(Intent.ACTION_VIEW,uri);
                startActivity(it);
                break;
            case R.id.button5:
                uri = Uri.parse("smsto:5556");
                it = new Intent(Intent.ACTION_SENDTO, uri);
                it.putExtra("sms_body", "The SMS text");
                startActivity(it);
                break;
            case R.id.button6:
                uri = Uri.parse("mailto:xxx@abc.com");
                it = new Intent(Intent.ACTION_SENDTO, uri);
                startActivity(it);
                break;
            case R.id.button7:
                uri = Uri.withAppendedPath(
                    MediaStore.Audio.Media.INTERNAL_CONTENT_URI, "1");
                it = new Intent(Intent.ACTION_VIEW, uri);
                startActivity(it);
                break;
        }
    }
}
```

7.7 同步练习二

继续 7.6 节中所介绍的例子，完善这个例子，要求如下。
（1）将显示在按钮右边的字体调大一些，以便和左边的按钮大小相匹配；
（2）编写自己的一个 Activity，作为图 7-7 界面中的一项功能能够启动并运行它，同时，该 Activity 将返回一些数据给主 Activity，主 Activity 能够显示返回的数据。

7.8 广播消息和广播接收器

Intent 不仅可以用于启动新的 Activity，也可以用于在应用程序之间广播消息。这一节我们就介绍如何使用 Intent 来广播消息，同时介绍如何接收 Android 平台的广播消息以便使应用程序能够对系统状态的变化做出响应。

Android 提供了两种可广播消息，即普通消息和有序消息。所谓普通消息，是指所有对该消息感兴趣的接收者都可以接收到该消息，并且它们接收到该消息的顺序是随机的；而有序消息则是按指定的优先级将消息先传递给具有最高优先级的接收者，具有最高优先级的接收者可以决定是否继续转发该消息给其他低优先级的接收者。这里我们只介绍普通消息，关于发送和接收有序消息及为消息设置接收权限的内容，请读者参见 Android SDK 文档。

7.8.1 发送和接收普通消息

任何 Activity，都可以使用 sendBroadcast(Intent intent)和 sendBroadcast(Intent intent, String receiverPermission)方法来发送普通广播消息。其中，sendBroadcast(Intent intent)方法将 Intent 消息发送给所有的接收者，而不管接收者是否具有接收权限；sendBroadcast(Intent intent, String receiverPermission)方法发送 Intent 消息给具有接收权限的接收者。

要接收 sendBroadcast 发送的消息，接收者必须继承 BroadcastReceiver 类，且需要实现其中的 onReceive(Context context, Intent intent)方法。并且我们需要在 AndroidManifest.xml 文件中或在程序代码中注册这个接收者。

下面我们举一个例子来说明如何发送和接收普通消息。在这个例子的首界面有一个输入框和一个按钮，我们在输入框中输入要发送的消息，点击按钮即可将消息发送出去。同时，消息接收者接收到该消息后，通过 Toast 将该消息显示出来。程序运行界面如图 7-10 所示。

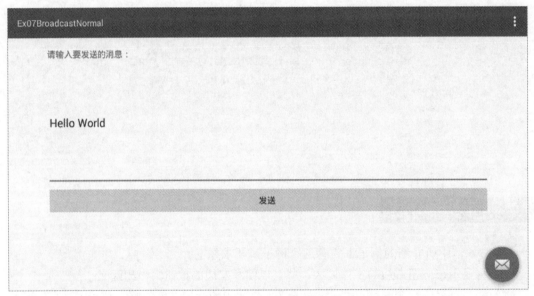

图 7-10　广播程序首界面

在如图 7-10 所示的输入框中输入一个字符串，然后点击发送按钮将消息发送出去，此时，将出现一个 Toast 来显示接收到的消息，如图 7-11 所示。

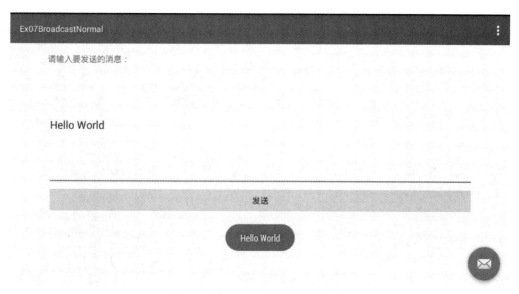

图 7-11　在 Toast 中显示接收到的消息

在 Android Studio 中新建一个名为 Ex07BroadcastNormal 的 Android 工程。

首先修改主界面布局文件，打开 res/layout/activity_main.xml 文件，将其中的内容替换为如下内容：

```xml
<?xml version="1.0" encoding="utf-8"?>
<LinearLayout xmlns:android="http://schemas.android.com/apk/res/android"
    android:layout_width="match_parent"
    android:layout_height="match_parent"
    android:orientation="vertical">

    <TextView
        android:layout_width="match_parent"
        android:layout_height="wrap_content"
        android:text="@string/text_input_something" />

    <EditText
        android:id="@+id/id_edittext"
        android:layout_width="match_parent"
        android:layout_height="200dp"
        android:singleLine="false"
        android:inputType="text"/>

    <Button
        android:id="@+id/id_button"
        android:layout_width="match_parent"
        android:layout_height="wrap_content"
        android:text="@string/text_button"/>

</LinearLayout>
```

这个布局文件很简单，只是在 LinearLayout 中放置了几个基本组件。

接着修改 res/values/strings.xml 文件，在其中添加布局文件中用到的几个字符串引用，修改后的文件内容如下：

```xml
<?xml version="1.0" encoding="utf-8"?>
<resources>

    <string name="app_name">Ex07BroadcastNormal</string>

    <string name="text_input_something">请输入要发送的消息：</string>
    <string name="text_button">发送</string>

</resources>
```

然后，再修改 MainActivity.java 文件，使其显示主界面，监听对按钮的点击事件，并在 onClick()方法中发送普通广播消息，修改后的代码如下：

```java
package com.ttt.ex07broadcastnormal;

import android.content.Intent;
import android.os.Bundle;
import android.support.v7.app.AppCompatActivity;
import android.view.View;
import android.widget.Button;
import android.widget.EditText;

public class MainActivity extends AppCompatActivity  implements View.OnClickListener {
    private EditText et;

    @Override
    protected void onCreate(Bundle savedInstanceState) {
        super.onCreate(savedInstanceState);
        setContentView(R.layout.activity_main);

        et = (EditText)this.findViewById(R.id.id_edittext);
        Button btn = (Button)this.findViewById(R.id.id_button);
        btn.setOnClickListener(this);

    }

    @Override
    public void onClick(View v) {
        String msg = et.getText().toString();

        Intent intent = new Intent("com.ttt.ex07broadcastnormal.hello");
        intent.putExtra("msg", msg);
        this.sendBroadcast(intent);
    }

}
```

注意其中的 onClick()方法的实现代码，在其中，我们创建一个 Intent 对象，并在该对象中保存相应的消息，然后广播该消息。

现在我们需要编写一个消息器来接收该消息。为此，在 src 目录下的 com.ttt.ex07broadcastnormal 包中新建一个名为 MyReceiver 的 Java 类，并修改其中的代码为如下内容：

```java
package com.ttt.ex07broadcastnormal;
```

```java
import android.content.BroadcastReceiver;
import android.content.Context;
import android.content.Intent;
import android.widget.Toast;

public class MyReceiver extends BroadcastReceiver {

    @Override
    public void onReceive(Context context, Intent intent) {
        String msg = intent.getStringExtra("msg");
        Toast.makeText(context, msg, Toast.LENGTH_LONG).show();
    }

}
```

作为消息接收者程序，其必须继承 BroadcastReceiver 类，并重写其中的 onReceive() 方法。在我们编写的接收者程序中，只是将接收到的消息通过 Toast 显示出来。

最后一步，需要在 AndroidManifest.xml 文件中注册该消息接收器，为此，修改 AndroidManifest.xml 文件为如下内容：

```xml
<?xml version="1.0" encoding="utf-8"?>
<manifest xmlns:android="http://schemas.android.com/apk/res/android"
    package="com.ttt.ex07broadcastnormal">

    <application
        android:allowBackup="true"
        android:icon="@mipmap/ic_launcher"
        android:label="@string/app_name"
        android:supportsRtl="true"
        android:theme="@style/AppTheme">
        <activity
            android:name=".MainActivity"
            android:label="@string/app_name"
            android:theme="@style/AppTheme.NoActionBar">
            <intent-filter>
                <action android:name="android.intent.action.MAIN" />

                <category android:name="android.intent.category.LAUNCHER" />
            </intent-filter>
        </activity>

        <receiver android:name=".MyReceiver">
            <intent-filter>
                <action android:name="com.ttt.ex07broadcastnormal.hello"/>
            </intent-filter>
        </receiver>

    </application>

</manifest>
```

注意其中的代码片段：

```xml
    <receiver android:name=".MyReceiver">
```

```xml
        <intent-filter>
            <action android:name="com.ttt.ex07broadcastnormal.hello"/>
        </intent-filter>
    </receiver>
```

这段代码注册了消息接收器 MyReceiver，使其可以监听 action 中 name 为 "com.ttt.ex07 broadcastnormal.hello" 的消息。

运行完成的程序，即可得到如图 7-10 所示的界面，在输入框中输入一些文字，点击"发送"按钮，即可将消息发送到 MyReceiver，并在 MyReceiver 中通过 Toast 将接收到的消息显示出来。

除了可以在 AndroidManifest.xml 文件中注册消息接收者（称为静态注册），我们还可以在 Java 代码中注册消息接收者（称为动态注册），例如，我们从 AndroidManifest.xml 文件中删除对 MyReceiver 的注册，并在 MainActivity.java 文件中来动态注册 MyReceiver，修改后的 MainActivity.java 代码如下：

```java
package com.ttt.ex07broadcastnormal;

import android.content.Intent;
import android.content.IntentFilter;
import android.os.Bundle;
import android.support.v7.app.AppCompatActivity;
import android.view.View;
import android.widget.Button;
import android.widget.EditText;

public class MainActivity extends AppCompatActivity  implements View.OnClickListener {
    private EditText et;
    private MyReceiver receiver;

    @Override
    protected void onCreate(Bundle savedInstanceState) {
        super.onCreate(savedInstanceState);
        setContentView(R.layout.activity_main);

        et = (EditText)this.findViewById(R.id.id_edittext);
        Button btn = (Button)this.findViewById(R.id.id_button);
        btn.setOnClickListener(this);

        receiver = new MyReceiver();
    }

    @Override
    public void onClick(View v) {
        String msg = et.getText().toString();

        Intent intent = new Intent("com.ttt.ex07broadcastnormal.hello");
        intent.putExtra("msg", msg);
        this.sendBroadcast(intent);
    }

    @Override
    protected void onResume() {
```

```java
        super.onResume();
        
        IntentFilter filter = new IntentFilter();
        filter.addAction("com.ttt.ex07broadcastnormal.hello");
        registerReceiver(receiver, filter);
    }

    @Override
    protected void onPause() {
        super.onPause();
        
        unregisterReceiver(receiver);
    }
}
```

在修改后的 MainActivity.java 文件中，我们在 onResume()生命周期方法中注册消息接收者 MyReceiver，并在 onPause()生命周期方法中解除对广播的监听。

运行修改后的程序，将得到与图 7-10 一致的效果。

7.8.2 接收 Android 平台的广播消息

Android 平台在特定的情况下会广播如下的普通消息，任何感兴趣的接收者通过注册均可以接收到这些广播消息。Android 平台常用的广播消息及其含义如表 7-1 所示。

表 7-1 Android 平台常用的广播消息及其含义

名　　称	含　　义
Intent.ACTION_TIME_TICK	每一分钟发送一次消息
Intent.ACTION_BOOT_COMPLETED	Android 完成启动时发送该消息
Intent.ACTION_SHUTDOWN	关机时发送该消息
Intent.ACTION_BATTERY_LOW	电池低于指定值时发送该消息
Intent.ACTION_POWER_CONNECTED	连接电源时发送该消息
Intent.ACTION_POWER_DISCONNECTED	断开电源连接时发送该消息

7.9 同步练习三

编写一个消息接收器程序，该程序既可以接收自己编写的程序发送的普通广播消息，也可以接收 Android 平台发送的 Intent.ACTION_POWER_CONNECTED 消息，并设计界面测试该程序。

菜单和 Toolbar

菜单是一种常见的应用程序操作模式，在早期的 Android 设备上，其提供了专门用于开启菜单的功能按钮，但是，从 Android 3.0 版本开始，Android 已经不再要求设备制造商提供这个功能按钮，而用 ActionBar 或 Toolbar 的应用程序组件来代替，这些组件在运行时出现在应用程序的最上方。虽然如此，ActionBar 及 Toolbar 组件仍是以基本 Android 的 Menu（菜单）组件来进行构造的。

8.1 菜单

通过 Android Studio 构建的基本程序的界面中，会显示一个菜单弹出按钮（黑框框住部分），如图 8-1 所示。

图 8-1　菜单弹出按钮

点击这个按钮，将显示菜单功能项，如图 8-2 所示。

第 8 章 菜单和 Toolbar

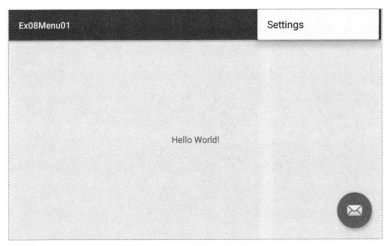

图 8-2 菜单功能项

菜单功能项中可以显示更多的选项。下面举一个例子来介绍 Android 菜单的使用方法。

在 Android Studio 中新建一个名为 Ex08Menu01 的工程，并修改 res/menu/menu_main.xml 文件为如下内容：

```xml
<menu xmlns:android="http://schemas.android.com/apk/res/android"
    xmlns:app="http://schemas.android.com/apk/res-auto"
    xmlns:tools="http://schemas.android.com/tools"
    tools:context="com.ttt.ex08menu01.MainActivity">

    <item
        android:id="@+id/id_mi_phone"
        android:icon="@drawable/png0030"
        android:orderInCategory="100"
        app:showAsAction="ifRoom|withText"
        android:title="@string/text_mi_phone"/>

    <item
        android:id="@+id/id_mi_cut"
        android:icon="@drawable/png0031"
        android:orderInCategory="100"
        app:showAsAction="ifRoom|withText"
        android:title="@string/text_mi_cut"/>

    <item
        android:id="@+id/id_mi_trump"
        android:icon="@drawable/png0032"
        android:orderInCategory="100"
        app:showAsAction="ifRoom|withText"
        android:title="@string/text_mi_trump"/>

    <item
        android:id="@+id/id_mi_book"
        android:icon="@drawable/png0033"
        android:orderInCategory="100"
        app:showAsAction="ifRoom|withText"
        android:title="@string/text_mi_book"/>
```

```xml
    <item
        android:id="@+id/id_mi_save"
        android:icon="@drawable/png0034"
        android:orderInCategory="99"
        app:showAsAction="ifRoom|withText"
        android:title="@string/text_mi_save"/>

    <item
        android:id="@+id/id_mi_mail"
        android:icon="@drawable/png0035"
        android:orderInCategory="100"
        app:showAsAction="ifRoom|withText"
        android:title="@string/text_mi_mail"/>

</menu>
```

并在 res/drawable 目录下放置需要的图片文件：png0030.png 到 png0035.png 文件。

在这个菜单中，共包括 6 个菜单项，并且我们为每个菜单项指定了 ID、图标和标题，同时，注意每个菜单项中的 android:orderInCategory，它指定将菜单项显示在 ActionBar 上的优先级，数值越小则优先级越高；app:showAsAction 指定了是否将该菜单项显示在 ActionBar 或 Toolbar 中，可取值为：never、ifRoom、withText 和 always，我们也可以将这些可取值进行或操作。当然，我们还需要修改 res/values/strings.xml 文件，在其中定义一些字符串引用，修改后的 res/values/strings.xml 文件内容如下：

```xml
<?xml version="1.0" encoding="utf-8"?>
<resources>

    <string name="app_name">Ex08Menu01</string>
    <string name="text_hello">ActionBar 示例</string>

    <string name="text_mi_phone">电话</string>
    <string name="text_mi_cut">剪切</string>
    <string name="text_mi_trump">喇叭</string>
    <string name="text_mi_book">书籍</string>
    <string name="text_mi_save">保存</string>
    <string name="text_mi_mail">邮件</string>

</resources>
```

为了能够在 ActionBar 中显示菜单，或用户在手机上按 MENU 按钮时显示菜单，还需要重写 MainActivity 的 onCreateOptionsMenu()方法，同时，为了处理对菜单项的点击事件，还需要重写 MainActivity 的 onOptionsItemSelected()方法，修改后的 MainActivity.java 文件如下：

```java
package com.ttt.ex08menu01;

import android.os.Bundle;
import android.support.design.widget.FloatingActionButton;
import android.support.design.widget.Snackbar;
import android.support.v7.app.AppCompatActivity;
import android.support.v7.widget.Toolbar;
import android.view.View;
import android.view.Menu;
import android.view.MenuItem;
```

```java
import android.widget.Toast;

public class MainActivity extends AppCompatActivity {

    @Override
    protected void onCreate(Bundle savedInstanceState) {
        super.onCreate(savedInstanceState);
        setContentView(R.layout.activity_main);
        Toolbar toolbar = (Toolbar) findViewById(R.id.toolbar);
        setSupportActionBar(toolbar);

        FloatingActionButton fab = (FloatingActionButton) findViewById(R.id.fab);
        fab.setOnClickListener(new View.OnClickListener() {
            @Override
            public void onClick(View view) {
                Snackbar.make(view, "Replace with your own action",
                        Snackbar.LENGTH_LONG)
                        .setAction("Action", null).show();
            }
        });
    }

    @Override
    public boolean onCreateOptionsMenu(Menu menu) {
        getMenuInflater().inflate(R.menu.menu_main, menu);
        return true;
    }

    @Override
    public boolean onOptionsItemSelected(MenuItem item) {
        int id = item.getItemId();
        switch(id) {
            case R.id.id_mi_phone:
                Toast.makeText(this, "启动打电话", Toast.LENGTH_LONG).show();
                break;
            case R.id.id_mi_cut:
                Toast.makeText(this, "剪切", Toast.LENGTH_LONG).show();
                break;
            case R.id.id_mi_trump:
                Toast.makeText(this, "吹喇叭", Toast.LENGTH_LONG).show();
                break;
            case R.id.id_mi_book:
                Toast.makeText(this, "书籍", Toast.LENGTH_LONG).show();
                break;
            case R.id.id_mi_save:
                Toast.makeText(this, "假装的保存", Toast.LENGTH_LONG).show();
                break;
            case R.id.id_mi_mail:
                Toast.makeText(this, "启动发邮件功能", Toast.LENGTH_LONG).show();
                break;
        }
        return super.onOptionsItemSelected(item);
    }
}
```

现在注意 onCreateOptionsMenu()方法。当用户点击手机上的 MENU 按钮，或者在系统生成 ActionBar 或 Toolbar 时将自动调用这个方法来生成所需要的菜单，因此，在这个方法中我们通过 getMenuInflater()获得一个菜单展开器并使用这个展开器来展开我们在 res/menu 目录下定义的 main.xml 菜单文件。

接着注意 onOptionsItemSelected()方法。这个方法是在用户点击菜单项时被调用的，当菜单列表被显示并且用户点击手机上的"返回"按钮时，也会调用这个方法，只是传递进来的参数的 ID 为 Android 内部的特定的 ID：android.R.id.home。我们正是通过这个 ID 来获知用户是否点击"返回"按钮的。

运行程序，将显示如图 8-3 所示的界面。

点击菜单上任意一个选项，将显示一个提示信息来表示菜单是工作的，当然，用户可以在菜单的功能代码中完成需要完成的任何工作。

图 8-3　菜单程序运行示例

8.2　ActionBar 和 Toolbar

观察图 8-3 中的程序运行结果最上面的那个组件，为了便于观察，将主要部分截出，如图 8-4 所示。

图 8-4　ActionBar 或 Toolbar

其中框住的部分称为 ActionBar 或 Toolbar。Android 建议使用 Toolbar 组件来作为菜单及程序标题的组件，因此，我们主要介绍 Toolbar 的使用方法。

我们查看 Android 的文档，会发现 Toolbar 组件是 ViewGroup 的子类，这说明，在 Toolbar 中可以嵌套任何 Android 的其他组件。不仅如此，Toolbar 还提供了相应方法来设置如 LOGO、Title、subtitle、UP 返回按钮等。

我们首先查看 MainActivity.java 程序的 onCreate()函数中的代码：

```java
public class MainActivity extends AppCompatActivity {

    @Override
    protected void onCreate(Bundle savedInstanceState) {
        super.onCreate(savedInstanceState);
        setContentView(R.layout.activity_main);
        Toolbar toolbar = (Toolbar) findViewById(R.id.toolbar);
        setSupportActionBar(toolbar);

        FloatingActionButton fab = (FloatingActionButton) findViewById(R.id.fab);
        fab.setOnClickListener(new View.OnClickListener() {
            @Override
            public void onClick(View view) {
                Snackbar.make(view, "Replace with your own action",
                        Snackbar.LENGTH_LONG)
                        .setAction("Action", null).show();
            }
        });
    }
    ……
}
```

读者注意这两句代码：

```java
Toolbar toolbar = (Toolbar) findViewById(R.id.toolbar);
setSupportActionBar(toolbar);
```

这两句代码就是获取界面中的 Toolbar 组件并将其作为 ActionBar 显示在界面的最上方。我们还可以在 Toolbar 组件上显示其他的信息，例如修改 MainActivity.java 的 onCreate()函数为如下内容：

```java
@Override
protected void onCreate(Bundle savedInstanceState) {
    super.onCreate(savedInstanceState);
    setContentView(R.layout.activity_main);
    Toolbar toolbar = (Toolbar) findViewById(R.id.toolbar);
    toolbar.setTitle("Toolbar");
    toolbar.setSubtitle("Sample");
    toolbar.setLogo(R.mipmap.ic_launcher);
    toolbar.setNavigationIcon(android.R.drawable.ic_media_rew);

    setSupportActionBar(toolbar);

    FloatingActionButton fab = (FloatingActionButton) findViewById(R.id.fab);
    fab.setOnClickListener(new View.OnClickListener() {
        @Override
        public void onClick(View view) {
            Snackbar.make(view, "Replace with your own action", Snackbar.LENGTH_LONG)
```

```
                .setAction("Action", null).show();
        }
    });
}
```

运行修改后的程序,在 ActionBar 上将显示更多的信息,如图 8-5 所示。

图 8-5 添加更多信息的 ActionBar

为了响应对最左边 按钮的点击事件,需要修改 MainActivity.java 文件的 onOptionsItemSelected()函数为如下内容:

```
@Override
public boolean onOptionsItemSelected(MenuItem item) {
    int id = item.getItemId();
    switch(id) {
        case R.id.id_mi_phone:
            Toast.makeText(this, "启动打电话", Toast.LENGTH_LONG).show();
            break;
        case R.id.id_mi_cut:
            Toast.makeText(this, "剪切", Toast.LENGTH_LONG).show();
            break;
        case R.id.id_mi_trump:
            Toast.makeText(this, "吹喇叭", Toast.LENGTH_LONG).show();
            break;
        case R.id.id_mi_book:
            Toast.makeText(this, "书籍", Toast.LENGTH_LONG).show();
            break;
        case R.id.id_mi_save:
            Toast.makeText(this, "假装的保存", Toast.LENGTH_LONG).show();
            break;
        case R.id.id_mi_mail:
            Toast.makeText(this, "启动发邮件功能", Toast.LENGTH_LONG).show();
            break;
        case android.R.id.home:
            Toast.makeText(this, "Home", Toast.LENGTH_LONG).show();
            return true;
    }
```

```
    return super.onOptionsItemSelected(item);
}
```

8.3 同步练习

探究练习：Toolbar 可以作为组件独立使用。现在编写一个简单的 Toolbar 练习程序，在 Toolbar 中显示两个按钮用于切换不同的图片：点击"上一张"按钮显示上一张图片，点击"下一张"按钮显示下一张图片，同时，在菜单上也实现同样的功能，并将菜单显示在 Toolbar 上。

动 画

Android 支持两种类型的动画：属性动画和 View 动画。View 动画又分为补间动画（Tween 动画）和帧动画（Frame 动画或 AnimationDrawable 动画）。动画的基本原理是在一个特定的时间内将组件的某个属性从一个值变化到一个新的值，或者将整个组件的显示状态从一种状态变化到一种新的状态。

View 动画是最常用的动画，因此，在本章我们主要对 View 动画的补间动画和帧动画进行介绍。

9.1 View 动画之补间动画基础

我们先介绍 View 动画的补间动画。先从一个简单的例子开始，读者对补间动画有一个初步认识后，我们再介绍补间动画的详细内容。

9.1.1 补间动画举例

所谓补间动画就是在一个特定的时间内，按一定的速度将组件从一个位置移动到另一个位置、从一个角度旋转到另一个角度、显示透明度从一个值变换到另一个值、大小从一个尺度变换到另一个尺度。不仅如此，Android 还允许将以上 4 种操作合并在一个动画中执行。

我们先看一个简单的例子，这个例子将一个 TextView 组件在 4 秒内以它的左下角为重心从 0 度顺时针旋转到 90 度，并且旋转的速度越来越快。运行效果如图 9-1 所示，期间，TextView 组件持续旋转，直到 4 秒后完成动画并显示最终结果。

第 9 章 动画

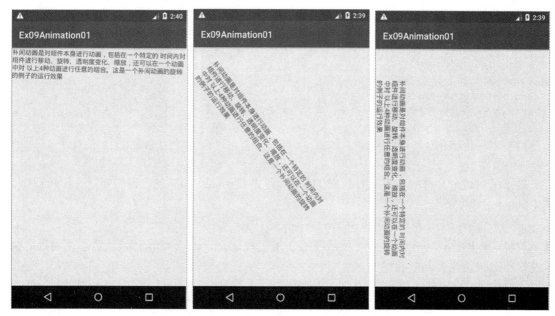

图 9-1 旋转 TextView 组件运行结果

现在我们来看一下这个例子的实现代码。首先新建一个名为 Ex09Animation01 的 Android 工程，然后查看布局文件 res/layout/activity_main.xml，其内容如下：

```xml
<RelativeLayout xmlns:android="http://schemas.android.com/apk/res/android"
    xmlns:tools="http://schemas.android.com/tools"
    android:layout_width="match_parent"
    android:layout_height="match_parent">

    <TextView
        android:id="@+id/id_tv"
        android:layout_width="wrap_content"
        android:layout_height="wrap_content"
        android:text="@string/hello_world" />

</RelativeLayout>
```

为了使 TextView 显示更多的文字，我们修改 res/values/strings.xml 文件内容为如下内容：

```xml
<?xml version="1.0" encoding="utf-8"?>
<resources>

    <string name="app_name">Ex09Animation01</string>
    <string name="hello_world">补间动画是对组件本身进行动画，包括在一个特定的
        时间内对组件进行移动、旋转、透明度变化、缩放，还可以在一个动画中对
        以上 4 种动画进行任意的组合。这是一个补间动画的旋转的例子的运行效果
    </string>

</resources>
```

与定义布局一样，Android 提供了两种定义动画的方式：通过 XML 文件定义动画和通过 Java 代码定义动画。Android 建议使用 XML 文件定义动画方式，并且 View 动画定义的文件必须放置在 res/anim 工程目录下。为此，我们首先在工程的 res 目录下创建 anim 目录，并在其目录下创建一个名为 my_rotate.xml 的动画文件。修改 my_rotate.xml 动画文件的内容为如下内容：

```xml
<?xml version="1.0" encoding="utf-8"?>
<rotate xmlns:android="http://schemas.android.com/apk/res/android"
    android:interpolator="@android:anim/accelerate_interpolator"
    android:fromDegrees="0"
    android:toDegrees="90"
    android:pivotX="0%"
    android:pivotY="100%"
    android:duration="4000"
    android:fillAfter="true">
</rotate>
```

接下来对这个文件的内容进行详细的解释。首先，<rotate>标签表示我们要定义旋转动画，其中的

```
android:interpolator="@android:anim/accelerate_interpolator"
```

表示旋转动画的执行方式，称为动画插值器，其值为 accelerate_interpolator，意思是动画的执行速度将越来越快。其中的

```
android:fromDegrees="0"
android:toDegrees="90"
```

表示旋转的初始角度和结束角度，分别为 0 度和 90 度。其中的

```
android:pivotX="0%"
android:pivotY="100%"
```

表示以组件的左下角为原点的旋转中心点的坐标。有两种表示中心点的方式：带%的值和不带%的值。带%的值表示以组件的长度或高度为度量基础，不带%的值表示以像素为度量基础。在我们不知道组件的实际大小的情况下，使用%会更容易一些，例如，在我们的例子中，我们指定旋转中心的 x 方向坐标偏移组件长度的 0%，也就是 x=0，而 y 方向坐标则偏移组件高度的 100%，也就是 y=组件的高度。其中的

```
android:duration="4000"
```

定义了动画执行的时间，单位是毫秒。其中的

```
android:fillAfter="true">
```

指定动画执行完成后组件的显示状态，true 表示组件保持动画执行完成后的状态，false 则表示组件在动画执行完成后，回到动画执行前的状态。

定义好动画文件后，接下来需要修改 MainActivity.java 文件，使 TextView 可以被动画处理，修改后的 MainActivity.java 文件内容如下：

```java
package com.ttt.ex09animation01;

import android.support.v7.app.AppCompatActivity;
import android.os.Bundle;
import android.view.animation.Animation;
import android.view.animation.AnimationUtils;
import android.widget.TextView;

public class MainActivity extends AppCompatActivity{

    @Override
    protected void onCreate(Bundle savedInstanceState) {
        super.onCreate(savedInstanceState);
        setContentView(R.layout.activity_main);
```

```
        TextView tv = (TextView) findViewById(R.id.id_tv);
        Animation rotate = AnimationUtils.loadAnimation(this, R.anim.my_rotate);
        tv.startAnimation(rotate);
    }
}
```

在 onCreate()回调函数中，我们获得 TextView 组件，并使用 AnimationUtils 类加载动画资源，然后在 TextView 组件上执行动画，内容如下：

```
        TextView tv = (TextView) findViewById(R.id.id_tv);
        Animation rotate = AnimationUtils.loadAnimation(this, R.anim.my_rotate);
        tv.startAnimation(rotate);
```

完成后运行该程序，即可得到如图 9-1 所示的效果。

9.1.2 补间动画类型

补间动画共包括 4 种动画形式，它们分别是：旋转动画、缩放动画、透明度动画和移位动画，标签分别为<rotate>、<scale>、<alpha>和<translate>。定义的补间动画的文件必须放置在 res/anim 工程目录下，文件名称可以是任何合法的名称，在程序中引用动画资源时，采用 R.anim.filename 的方式来引用，在 XML 文件中采用@[package:]anim/filename 的方式来引用动画资源。

1. 定义 rotate 旋转动画

定义 rotate 旋转动画的一般形式如下：

```
<?xml version="1.0" encoding="utf-8"?>
<rotate xmlns:android="http://schemas.android.com/apk/res/android"
    android:interpolator="@[package:]anim/interpolator_resource"
    android:fromDegrees="float"
    android:toDegrees="float"
    android:pivotX="float"
    android:pivotY="float"
    android:startOffset="int"
    android:duration="int"
    android:fillAfter="boolean"
    android:repeatCount="int"
    android:repeatMode="restart | reverse"
    >
</rotate>
```

其中的

```
    android:interpolator="@[package:]anim/interpolator_resource"
```

表示动画插值器，也就是采用何种方式来改变组件的值或组件的显示状态，Android 内置的动画插值器如图 9-2 所示。

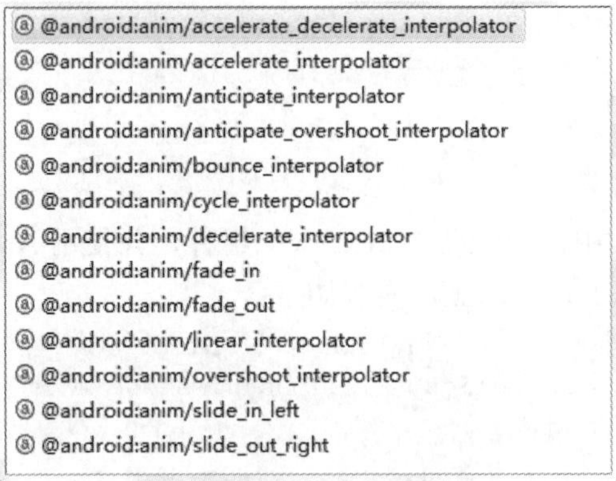

```
@android:anim/accelerate_decelerate_interpolator
@android:anim/accelerate_interpolator
@android:anim/anticipate_interpolator
@android:anim/anticipate_overshoot_interpolator
@android:anim/bounce_interpolator
@android:anim/cycle_interpolator
@android:anim/decelerate_interpolator
@android:anim/fade_in
@android:anim/fade_out
@android:anim/linear_interpolator
@android:anim/overshoot_interpolator
@android:anim/slide_in_left
@android:anim/slide_out_right
```

图 9-2　Android 内置的动画插值器

每个插值器的表现形式很难用语言表达清楚，读者可以在 9.1.1 节的例子中使用如图 9-2 所示的任意插值器替换例子中的 accelerate_interpolator 来查看各个插值器的表现形式。

动画定义中的

```
android:fromDegrees="float"
android:toDegrees="float"
android:pivotX="float"
android:pivotY="float"
```

表示旋转的初始角度、最终角度和旋转中心点，这些参数已经在 9.1.1 节中介绍过。其中的

```
android:startOffset="int"
android:duration="int"
```

分别表示启动动画的时间延迟和时间长度。其中的

```
android:fillAfter="boolean"
```

表示动画执行后是否在组件上保持动画执行后的最终结果。其中的

```
android:repeatCount="int"
android:repeatMode="int"
```

分别表示动画重复的次数和重复的方式，重复的方式可取值为：restart、reverse。顾名思义，restart 表示再次执行动画，reverse 表示以相反的方式再次执行动画。

2. 定义 scale 缩放动画

定义 scale 缩放动画的一般形式如下：

```xml
<?xml version="1.0" encoding="utf-8"?>
<scale xmlns:android="http://schemas.android.com/apk/res/android"
    android:interpolator="@[package:]anim/interpolator_resource"
    android:fromXScale="float"
    android:toXScale="float"
    android:fromYScale="float"
    android:toYScale="float"
    android:pivotX="float"
    android:pivotY="float"
    android:startOffset="int"
    android:duration="int"
    android:fillAfter="boolean"
    android:repeatCount="int"
    android:repeatMode="restart | reverse"
    >
```

```
</scale>
```
其中的
```
    android:fromXScale="float"
    android:toXScale="float"
```
表示在 x 方向的以组件的宽度为基础的组件宽度变化的比例,例如,android:fromXScale="1.4"、android:toXScale="0.5"表示宽度从 1.4 倍变换到 0.5 倍。其中的
```
    android:fromYScale="float"
    android:toYScale="float"
```
表示组件在 y 方向上的以组件的高度为基础的组件高度变化的比例。其他的参数与 rotate 中的解释相同。

3. 定义 alpha 透明度动画

定义 alpha 透明度动画的一般形式如下:
```
<?xml version="1.0" encoding="utf-8"?>
<alpha xmlns:android="http://schemas.android.com/apk/res/android"
    android:interpolator="@[package:]anim/interpolator_resource"
    android:fromAlpha="float"
    android:toAlpha="float"
    android:startOffset="int"
    android:duration="int"
    android:fillAfter="boolean"
    android:repeatCount="int"
    android:repeatMode="restart | reverse"
    >
</alpha>
```
其中的
```
    android:fromAlpha="float"
    android:toAlpha="float"
```
表示组件的透明度值的变化,取值从 0.0 到 1.0,其中,0.0 表示完全透明,1.0 表示完全不透明。其他参数与 rotate 中的解释相同。

4. 定义 translate 移位动画

定义 translate 移位动画的一般形式如下:
```
<?xml version="1.0" encoding="utf-8"?>
<translate xmlns:android="http://schemas.android.com/apk/res/android"
    android:interpolator="@[package:]anim/interpolator_resource"
    android:fromXDelta="float"
    android:fromYDelta="float"
    android:toXDelta="float"
    android:toYDelta="float"
    android:startOffset="int"
    android:duration="int"
    android:fillAfter="boolean"
    android:repeatCount="int"
    android:repeatMode="restart | reverse"
    >
</translate>
```
其中的

```
android:fromXDelta="float"
android:fromYDelta="float"
```

表示以组件的左上角为移动起始原点的坐标,有 3 种表示方式:普通数值、带%的数值和带%p的数值。其中普通数值表示像素点坐标值,例如,android:fromXDelta="5",表示起始点的 *x* 坐标为 5;带%的数值表示以组件的宽度或高度为度量基础的像素点的坐标,例如,android:fromXDelta="5%"表示起始点的 *x* 方向坐标为组件宽度的 5%;带%p 的数值表示以组件的父容器的宽度或高度为度量基础的像素点的坐标。其中的

```
android:toXDelta="float"
android:toYDelta="float"
```

表示移位结束时左上角所在的坐标点。表示方法与移位起始点相同。

5. 定义复合 set 动画

除了可以使用<rotate>、<scale>、<alpha>和<translate>标签定义单个动画资源,我们还可以使用<set>标签定义复合动画,所谓复合动画就是将多个动画合并成一个动画来执行。定义复合 set 动画的一般形式如下:

```
<?xml version="1.0" encoding="utf-8"?>
<set xmlns:android="http://schemas.android.com/apk/res/android"
      android:interpolator="@[package:]anim/interpolator_resource"
      android:shareInterpolator=["true" | "false"] >
   <rotate ……/>
   <translate ……/>
   ……
   <set ……> …… </set>
</set>
```

也就是说,set 动画是其他 4 类动画的组合,并且在 set 动画中还可以包含 set 动画。读者注意其中的一个配置参数:

```
android:shareInterpolator=["true" | "false"]>
```

表示在 set 中所指明的插值器是否作用于所有在 set 中定义的子动画,true 表示是,false 表示否。

9.1.3 使用动画监听器

我们可以使用动画 Animation.AnimationListener 监听器接口来监听动画执行的各个阶段,从 Android 的文档可以知道,该监听器接口包括 3 个方法,如图 9-3 所示。

Public Methods	
abstract void	onAnimationEnd (Animation animation) Notifies the end of the animation.
abstract void	onAnimationRepeat (Animation animation) Notifies the repetition of the animation.
abstract void	onAnimationStart (Animation animation) Notifies the start of the animation.

图 9-3 动画监听器接口的 3 个方法

这 3 个方法分别用于监听动画的开始(onAnimationStart)、结束(onAnimationEnd)和重复(onAnimationRepeat)。

我们可以修改 9.1.1 节中的旋转 TextView 的动画例子,动画结束时使用 Toast 显示一个结束消息。为此,我们只需要修改 MainActivity.java 程序的 onCreate()回调函数,修改后的 onCreate()

函数如下：
```
@Override
protected void onCreate(Bundle savedInstanceState) {
    super.onCreate(savedInstanceState);
    setContentView(R.layout.activity_main);

    TextView tv = (TextView) findViewById(R.id.id_tv);
    Animation rotate = AnimationUtils.loadAnimation(this, R.anim.my_rotate);
    rotate.setAnimationListener(new AnimationListener() {

        @Override
        public void onAnimationStart(Animation animation) {
        }

        @Override
        public void onAnimationEnd(Animation animation) {
            Toast.makeText(MainActivity.this, "动画结束",
                                Toast.LENGTH_LONG).show();;
        }

        @Override
        public void onAnimationRepeat(Animation animation) {
        }

    });
    tv.startAnimation(rotate);
}
```

在这个函数中，我们调用 rotate.setAnimationListener 设置 rotate 动画的监听器，并在 onAnimationEnd()方法中显示结束消息。运行修改后的程序，当动画结束时显示如图 9-4 所示的界面。

图 9-4 动画结束时的显示界面

9.2 View 动画之帧动画

帧动画就是按一定的时间顺序显示一组预定义的图片。采用 XML 文件定义帧动画时，需要将定义文件放置在 res/drawable 工程目录下，文件名可以是任意合法的文件名。在 XML 或在 Java 程序中引用帧动画资源与引用一般图片资源的形式是一样的，可以在任何可以使用 drawable 资源的地方使用帧动画资源。定义帧动画的一般格式如下：

```xml
<?xml version="1.0" encoding="utf-8"?>
<animation-list
    xmlns:android="http://schemas.android.com/apk/res/android"
    android:oneshot=["true" | "false"] >
      <item
          android:drawable="@[package:]drawable/drawable_resource_name"
                                              android:duration="integer" />
      <item ……/>
      ……
</animation-list>
```

其中的

```
android:oneshot=["true" | "false"]>
```

表示是循环执行动画还是一次性执行动画。其中的

```xml
<item
    android:drawable="@[package:]drawable/drawable_resource_name"
                                        android:duration="integer" />
    <item ……/>
    ……
```

用于定义在动画中要显示的图片，android:drawable 指明要显示的图片，android:duration 指明该图片显示的时间。

我们通过一个简单的例子来看一下如何在程序中使用帧动画。该例子将以每张图片显示 200 毫秒的时间顺序来显示如图 9-5 所示的图片。

图 9-5　帧动画显示的一组图片

该程序运行效果如图 9-6 所示。

第 9 章 动画

图 9-6 帧动画运行效果

现在我们来看一下程序的实现代码。新建一个名为 Ex09Animation02 的 Android 工程。在 res/drawable 目录下放置如图 9-5 所示的图片,并新建一个名为 drawable_rotate.xml 的帧动画资源文件。其中,drawable_rotate.xml 文件内容如下:

```xml
<?xml version="1.0" encoding="utf-8"?>
<animation-list xmlns:android="http://schemas.android.com/apk/res/android"
    android:oneshot="false">
    <item android:drawable="@drawable/rotate1" android:duration="200" />
    <item android:drawable="@drawable/rotate2" android:duration="200" />
    <item android:drawable="@drawable/rotate3" android:duration="200" />
    <item android:drawable="@drawable/rotate4" android:duration="200" />
</animation-list>
```

在这个帧动画资源文件中,程序顺序地显示 4 张图片,并设置每张图片显示时间为 200 毫秒。

现在修改 res/layout/activity_main.xml 文件,该布局在 LinearLayout 中显示一个 ImageView 组件和两个按钮,res/layout/activity_main.xml 文件内容如下:

```xml
<LinearLayout xmlns:android="http://schemas.android.com/apk/res/android"
    xmlns:tools="http://schemas.android.com/tools"
    android:layout_width="match_parent"
    android:layout_height="match_parent"
    android:paddingBottom="@dimen/activity_vertical_margin"
    android:paddingLeft="@dimen/activity_horizontal_margin"
    android:paddingRight="@dimen/activity_horizontal_margin"
    android:paddingTop="@dimen/activity_vertical_margin"
    tools:context="com.ttt.ex15animation03.MainActivity"
    android:orientation="vertical">

    <ImageView
        android:id="@+id/id_iamge_view"
        android:layout_width="wrap_content"
        android:layout_height="0dp"
        android:layout_weight="8"
        android:scaleType="fitCenter"
        android:src="@drawable/drawable_rotate"
```

145

```xml
        android:contentDescription="@string/hello_world"/>

    <Button
        android:id="@+id/id_button_1"
        android:layout_width="match_parent"
        android:layout_height="0dp"
        android:layout_weight="1"
        android:text="@string/text_start"
    />

    <Button
        android:id="@+id/id_button_2"
        android:layout_width="match_parent"
        android:layout_height="0dp"
        android:layout_weight="1"
        android:text="@string/text_stop"
    />

</LinearLayout>
```

同时，需要修改 res/values/strings.xml 文件，在其中定义在布局文件中用到的字符串引用，内容如下：

```xml
<?xml version="1.0" encoding="utf-8"?>
<resources>

    <string name="app_name">Ex09Animation02</string>
    <string name="hello_world">Hello world!</string>
    <string name="action_settings">Settings</string>

    <string name="text_start">开始</string>
    <string name="text_stop">停止</string>

</resources>
```

最后修改 MainActivity.java 文件，在点击按钮时启动或停止动画，修改后的文件内容如下：

```java
package com.ttt.ex09animation02;

import android.support.v7.app.AppCompatActivity;
import android.graphics.drawable.AnimationDrawable;
import android.os.Bundle;
import android.view.View;
import android.view.View.OnClickListener;
import android.widget.Button;
import android.widget.ImageView;

public class MainActivity extends AppCompatActivityimplements OnClickListener{
    private Button btn1, btn2;


    @Override
    protected void onCreate(Bundle savedInstanceState) {
        super.onCreate(savedInstanceState);
```

```java
        setContentView(R.layout.activity_main);

        btn1 = (Button)this.findViewById(R.id.id_button_1);
        btn1.setOnClickListener(this);
        btn2 = (Button)this.findViewById(R.id.id_button_2);
        btn2.setOnClickListener(this);
    }

    @Override
    public void onClick(View v) {
        int id = v.getId();
        if (id == R.id.id_button_1) {
            ImageView iv = (ImageView)this.findViewById(R.id.id_iamge_view);
            AnimationDrawable ad = (AnimationDrawable)iv.getDrawable();
            ad.start();
        }
        else {
            ImageView iv = (ImageView)this.findViewById(R.id.id_iamge_view);
            AnimationDrawable ad = (AnimationDrawable)iv.getDrawable();
            ad.stop();
        }
    }
}
```

在对按钮点击事件的处理中，我们从界面中获得 ImageView 的引用，并得到在其中显示的 Drawable 对象，由于 ImageView 的 Drawable 对象是一个帧动画，因此，得到该帧动画资源后，可以通过 start() 或 stop() 方法启动或停止动画的执行。

执行该程序，将得到如图 9-6 所示的结果，点击"开始"或"停止"按钮，将启动或停止该动画。

9.3 同步练习

编写一个简单的动画程序，在程序启动时首先执行一个开启动画，在动画执行完毕后再显示应用程序的主界面。例如，在启动程序时，读者可以在一个 ImageView 上执行一个动画，在动画执行完毕后显示程序的主界面。

多媒体播放

Android 的多媒体框架提供了播放音频、视频及图像的相关方式。通过该框架，我们可以处理来自资源文件、手机本地存储系统，甚至来自网络的多媒体内容。

Android 的多媒体框架中最重要和常用的类无疑就是 MediaPlayer，通过使用 MediaPlayer 对象，程序可以获取、解码和播放包括音频和视频在内的多媒体资源。所播放的媒体内容可以来自程序资源（放置在 res/raw 目录下，因为 Android 不会对放置在这个目录下的资源进行任何处理，所以，放置在这个目录下的文件将保持原样）、来自本地文件或来自网络等。

本章对使用 MediaPlayer 对象播放音频和视频的内容进行介绍。

10.1 使用 MediaPlayer 播放音频

对于需要在程序中播放的音频，大致可以分为两类：为程序增加音效的简短音频和播放任何来自本地文件系统及网络的音频。对于为增加程序音效而播放的简短音频，读者可以使用 MediaPlayer 提供的便利的方法进行播放；而对于播放来自本地文件或网络的音频，笔者建议使用 MediaPlayer 提供的完整播放方式进行播放。

10.1.1 播放简短的音频

为了播放简短的音频，MediaPlayer 提供了几种静态的 create()方法用于创建 MediaPlayer 对象。从 Android 的帮助文档中可以看出，MediaPlayer 提供的便利的 create()方法如图 10-1 所示。

static MediaPlayer	create(Context context, int resid) Convenience method to create a MediaPlayer for a given resource id.
static MediaPlayer	create(Context context, Uri uri) Convenience method to create a MediaPlayer for a given Uri.

图 10-1　MediaPlayer 提供的便利的 create()方法

其中，create(Context context, int resid)用于播放存放在 res/raw 资源目录下的音频资源，而 create(Context context, Uri uri)可以播放来自任何 Uri 的音频。对于需要播放的简短音频，笔者建议将音频文件存放在资源目录下，也就是 res/raw 目录下。

下面我们通过一个具体的例子来看看如何使用这种方式来播放简短的音频。该程序运行后，显示一个简短的时间长度为 4 秒的启动动画，期间将同步循环播放一段音频（因为音频的长度可能少于 4 秒，所以需要循环播放），动画结束时，我们也停止对音频的播放，运行效果如图 10-2 所示。

图 10-2　动画执行时播放简短音频

现在我们来看一下该程序的实现代码。为此，新建一个名为 Ex10MultiMedia01 的 Android 工程，并在 res 工程目录下分别新建名为 raw 和 anim 的子资源目录。在 res/raw 工程目录下放置一个要播放的音频文件，其文件名为 short4.mp3。现在来修改布局文件 res/layout/activity_main.xml，其内容如下：

```xml
<RelativeLayout xmlns:android="http://schemas.android.com/apk/res/android"
    android:layout_width="match_parent"
    android:layout_height="match_parent">

    <ImageView
        android:id="@+id/id_image_view"
        android:layout_width="match_parent"
        android:layout_height="match_parent"
        android:scaleType="fitCenter"
        android:src="@drawable/ic_launcher"
        android:contentDescription="@string/hello_world" />

</RelativeLayout>
```

这个布局文件很简单，只是显示一个 ImageView 而已。然后，在 res/anim 目录下新建一个名为 my_scale.xml 的动画文件，其内容如下：

```xml
<?xml version="1.0" encoding="utf-8"?>
<scale xmlns:android="http://schemas.android.com/apk/res/android"
    android:interpolator="@android:anim/linear_interpolator"
    android:fromXScale="0"
    android:toXScale="1"
    android:fromYScale="0"
    android:toYScale="1"
    android:pivotX="50%"
    android:pivotY="50%"
    android:duration="4000"
    android:fillAfter="true"
    >
</scale>
```

这个动画文件将以其中心点为中心执行这个动画的组件，在 4 秒内，将组件在 x 方向和 y 方向从 0.0 放大到组件的正常尺度。最后修改 MainActivity.java 文件，修改后的内容如下：

```java
package com.ttt.ex10multimedia01;

import android.support.v7.app.AppCompatActivity;
import android.media.MediaPlayer;
import android.os.Bundle;
import android.view.animation.Animation;
import android.view.animation.AnimationUtils;
import android.view.animation.Animation.AnimationListener;
import android.widget.ImageView;

public class MainActivity extends AppCompatActivity {
    private MediaPlayer mp;

    @Override
    protected void onCreate(Bundle savedInstanceState) {
        super.onCreate(savedInstanceState);
        setContentView(R.layout.activity_main);

        mp = null;

        ImageView iv = (ImageView) findViewById(R.id.id_image_view);
        Animation rotate = AnimationUtils.loadAnimation(this, R.anim.my_scale);
        rotate.setAnimationListener(new AnimationListener() {

            @Override
            public void onAnimationStart(Animation animation) {
                mp = MediaPlayer.create(MainActivity.this, R.raw.short4);
                mp.setLooping(true);
                mp.setVolume(0.5f, 0.5f);
                mp.start();
            }

            @Override
            public void onAnimationEnd(Animation animation) {
                mp.stop();
                mp.release();
                mp = null;
            }

            @Override
            public void onAnimationRepeat(Animation animation) {

            }

        });
        iv.startAnimation(rotate);
    }
}
```

主要工程都是在该类的 onCreate() 函数中完成的。在该函数中，首先获得 ImageView 的引用，加载动画资源，并设置动画执行过程监听器。在监听器的 onAnimationStart() 函数中，我们通过如下代码来启动播放一个短音频：

```
mp = MediaPlayer.create(MainActivity.this, R.raw.short4);
mp.setLooping(true);
mp.setVolume(0.5f, 0.5f);
mp.start();
```

其中，我们通过 MediaPlayer.create()方法创建 MediaPlayer 对象，在这里，我们加载并解码了本地音频资源，需要说明的是，对于存储从网络上下载的音频资源，MediaPlayer.create()可能需要较长的时间进行加载和解码，因此,当在 Android 生成的 UI 线程中执行 MediaPlayer.create()方法时,可能出现"响应超时"错误。在加载并解码了音频资源后,我们调用 MediaPlayer 的 setLooping()方法设置循环播放该音频，调用 setVolume()方法设置该音频的播放音量，最后调用 start()方法播放音频。在监听器的 onAnimationEnd()方法中，我们通过代码：

```
mp.stop();
mp.release();
mp = null;
```

停止播放音频，并释放 MediaPlayer 对象所占用的系统资源，注意，在必须要关闭播放器的时候，一定要调用其 release()方法释放其占用的资源。

现在运行这个程序，将显示如图 10-2 所示的界面，在 ImageView 逐渐放大的同时，将播放一段音频。

10.1.2 使用 MediaPlayer 自制一个音频播放器

我们通过编写一个播放本地 SD 卡上的音频文件的播放器程序来介绍如何使用 MediaPlayer 提供的完整播放方式来播放音频文件。该程序首先显示 SD 卡上的目录及文件列表。程序运行效果如图 10-3 所示。

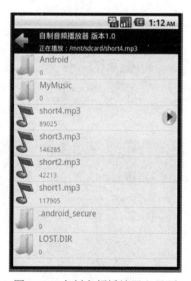

图 10-3　自制音频播放器主界面

此时，我们可以点击任何一个文件，程序将播放所点击的文件，并将在列表的右边显示一个播放图标表示正在播放此文件，同时，会在界面上方的"正在播放"文本框中显示正在播放的文件。我们还可以点击目录进入子目录中，点击界面左上角的箭头返回到上一级目录。

现在我们来看一下该程序的实现代码。新建一个名为 Ex10MultiMedia02 的 Android 工程，工程创建完成后，在 res/drawable 目录下放置需要用到的图标，包括文件浏览图标（界面左上角的箭头)、文件类型图标和播放图标。

我们来看一下主界面布局文件 res/layout/activity_main.xml，该布局文件的内容如下：

```xml
<LinearLayout xmlns:android="http://schemas.android.com/apk/res/android"
    android:layout_width="match_parent"
    android:layout_height="match_parent"
    android:orientation="vertical">

    <LinearLayout
        android:layout_width="match_parent"
        android:layout_height="48dp"
        android:background="#FF000000"
        android:orientation="horizontal" >

        <ImageButton
            android:id="@+id/id_image_button"
            android:layout_width="32dp"
            android:layout_height="32dp"
            android:layout_gravity="center"
            android:background="@drawable/png1638"
            android:scaleType="fitCenter"
            android:contentDescription="@string/hello_world"
        />

        <View
            android:layout_width="16dp"
            android:layout_height="match_parent"
        />

        <LinearLayout
            android:layout_width="match_parent"
            android:layout_height="match_parent"
            android:orientation="vertical"
        >

            <TextView
                android:id="@+id/id_text_view_1"
                android:layout_width="match_parent"
                android:layout_height="28dp"
                android:gravity="center_vertical"
                android:textSize="14sp"
                android:textColor="#FFFFFFFF"
                android:text="@string/text_version"
            />

            <TextView
                android:id="@+id/id_text_view_2"
                android:layout_width="match_parent"
                android:layout_height="20dp"
                android:gravity="center_vertical"
                android:textSize="12sp"
                android:textColor="#FFFFFFFF"
                android:text="@string/text_current_playing"
```

```
            />
        </LinearLayout>
    </LinearLayout>
    <ListView
        android:id="@+id/id_list_view"
        android:layout_width="match_parent"
        android:layout_height="match_parent"
    />
</LinearLayout>
```

该布局文件显示如图 10-3 所示的界面布局：在 ListView 的上面显示目录浏览图标、软件名称和版本、正在播放的文件。由于在该布局文件中使用到了一些字符串引用，因此，需要修改 res/values/strings.xml 文件，其内容如下：

```xml
<?xml version="1.0" encoding="utf-8"?>
<resources>

    <string name="app_name">Ex10MultiMedia02</string>
    <string name="hello_world">Hello world!</string>

    <string name="text_version">自制音频播放器 版本 1.0</string>
    <string name="text_current_playing">正在播放：</string>

</resources>
```

现在来看一下 ListView 的列表项的布局文件 res/layout/list_item.xml，其内容如下：

```xml
<?xml version="1.0" encoding="utf-8"?>
<RelativeLayout xmlns:android="http://schemas.android.com/apk/res/android"
    android:layout_width="match_parent"
    android:layout_height="48dp"
    android:layout_gravity="center_vertical"
    >

    <ImageView
        android:id="@+id/id_li_image_view_type"
        android:layout_width="44dp"
        android:layout_height="44dp"
        android:layout_alignParentLeft="true"
        android:scaleType="fitCenter"
        android:contentDescription="@string/hello_world"
    />

    <View
        android:id="@+id/id_li_space_1"
        android:layout_width="6dp"
        android:layout_height="44dp"
        android:layout_toRightOf="@id/id_li_image_view_type"
    />

    <LinearLayout
```

```xml
        android:layout_width="wrap_content"
        android:layout_height="44dp"
        android:layout_toRightOf="@id/id_li_space_1"
        android:orientation="vertical" >

        <TextView
            android:id="@+id/id_li_text_view_name"
            android:layout_width="wrap_content"
            android:layout_height="28dp"
            android:textSize="16sp"
        />

        <TextView
            android:id="@+id/id_li_text_view_length"
            android:layout_width="wrap_content"
            android:layout_height="16dp"
            android:textSize="12sp"
        />

    </LinearLayout>

    <ImageView
        android:id="@+id/id_li_image_view_play"
        android:layout_width="32dp"
        android:layout_height="32dp"
        android:layout_alignParentRight="true"
        android:layout_centerVertical="true"
        android:scaleType="fitCenter"
        android:contentDescription="@string/hello_world"
    />

</RelativeLayout>
```

在列表项中，界面显示文件类型图标、文件名、文件长度及播放的图标，如图 10-4 所示。

图 10-4 列表项布局示例

为了在 ListView 组件显示文件列表，需要构建一个 Adapter 来为 ListView 提供数据，为此，在 src 工程目录下，新建一个名为 com.ttt.ex10multimedia02.adapters 的包，并在这个包下新建一个名为 FileListAdapter.java 的文件，该文件的内容如下：

```java
package com.ttt.ex10multimedia02.adapters;

import java.io.File;
import java.util.ArrayList;

import com.ttt.ex10multimedia02.R;

import android.content.Context;
import android.os.Environment;
```

```java
import android.view.LayoutInflater;
import android.view.View;
import android.view.ViewGroup;
import android.widget.BaseAdapter;
import android.widget.ImageView;
import android.widget.ListView;
import android.widget.RelativeLayout;
import android.widget.TextView;

public class FileListAdapter extends BaseAdapter {
    private File cFile;
    private ArrayList<FileDesc> cList;

    private String currentPlaying;

    LayoutInflater inflater;

    public FileListAdapter(Context context) {
        inflater = (LayoutInflater)
                    context.getSystemService(Context.LAYOUT_INFLATER_SERVICE);
        currentPlaying = null;

        cList = new ArrayList<FileDesc>();
        cFile = Environment.getExternalStorageDirectory();

        load();
    }

    @Override
    public int getCount() {
        return cList.size();
    }

    @Override
    public Object getItem(int position) {
        return null;
    }

    @Override
    public long getItemId(int position) {
        return 0;
    }

    @Override
    public View getView(int position, View convertView, ViewGroup parent) {
        FileDesc fd = cList.get(position);
        RelativeLayout rl;

        if (convertView != null)
            rl = (RelativeLayout)convertView;
        else
            rl = (RelativeLayout)inflater.inflate(R.layout.list_item, parent, false);
```

```java
        rl.setTag(fd.file.getAbsolutePath());

        ImageView iv_type = (ImageView)rl.findViewById(R.id.id_li_image_view_type);
        if (fd.type.equalsIgnoreCase("dir"))
            iv_type.setImageResource(R.drawable.png1674);
        else
            iv_type.setImageResource(R.drawable.png0088);

        TextView tv_name = (TextView)rl.findViewById(R.id.id_li_text_view_name);
        tv_name.setText(fd.name);

        TextView tv_length = (TextView)rl.findViewById(R.id.id_li_text_view_length);
        tv_length.setText(fd.length + "");

        ImageView iv_play = (ImageView)rl.findViewById(R.id.id_li_image_view_play);
        iv_play.setImageResource(R.drawable.png1702);
        if ((fd.file.getAbsolutePath()).equalsIgnoreCase(currentPlaying))
            iv_play.setVisibility(View.VISIBLE);
        else
            iv_play.setVisibility(View.INVISIBLE);

        return rl;
    }

    private void load() {
        File[] fl = cFile.listFiles();
        if (fl == null)
            return;

        cList.clear();

        int count = fl.length;
        for(int i=0; i<count; i++) {
            FileDesc fd = new FileDesc();
            fd.file = fl[i];
            fd.name = fl[i].getName();
            fd.length = fl[i].length();
            fd.type = fl[i].isDirectory()==true?"dir":"file";

            cList.add(fd);
        }

        this.notifyDataSetChanged();
    }

    public void goUp() {
        if (Environment.getExternalStorageDirectory().getAbsolutePath().
                            equalsIgnoreCase(cFile.getAbsolutePath())) {
            return;
        }

        cFile = cFile.getParentFile();
        load();
```

```java
    }

    public void goDown(File ndir) {
        cFile = ndir;
        load();
    }

    public String getType(int position) {
        return cList.get(position).type;
    }

    public File getFile(int position) {
        return cList.get(position).file;
    }

    public File setAndPlay(ListView lv, int position) {
        FileDesc fd = cList.get(position);
        RelativeLayout rl;

        if (currentPlaying != null) {
            rl = (RelativeLayout)lv.findViewWithTag(currentPlaying);
            if (rl != null) {
                ImageView iv_play =
                        (ImageView)rl.findViewById(R.id.id_li_image_view_play);
                iv_play.setVisibility(View.INVISIBLE);
            }
        }

        currentPlaying = fd.file.getAbsolutePath();
        rl = (RelativeLayout)lv.findViewWithTag(currentPlaying);
        if (rl != null) {
            ImageView iv_play = (ImageView)rl.findViewById(R.id.id_li_image_view_play);
            iv_play.setVisibility(View.VISIBLE);
        }

        return cList.get(position).file;
    }

    private class FileDesc {
        public File file;
        public String name;
        public long length;
        public String type;
    }
}
```

我们先来看其中的变量定义，内容如下：

```java
private File cFile;
private ArrayList<FileDesc> cList;

private String currentPlaying;
```

其中的 cFile 表示当前正在主界面显示的目录 File 对象，cList 表示当前目录的文件列表，其中的 FileDesc 是一个自定义的类，用于表示我们想要展示的文件信息，FileDesc 的定义如下：

```
private class FileDesc {
    public File file;
    public String name;
    public long length;
    public String type;
}
```

我们只想要展示文件的文件名、长度和类型，同时，为了便于后续操作，我们将每个文件的 File 对象也保存在 FileDesc 对象中。我们再来看构造函数：

```
public FileListAdapter(Context context) {
    inflater = (LayoutInflater)
                context.getSystemService(Context.LAYOUT_INFLATER_SERVICE);
    currentPlaying = null;

    cList = new ArrayList<FileDesc>();
    cFile = Environment.getExternalStorageDirectory();

    load();
}
```

在构造函数中，我们创建 cList 对象，并将当前目录 cFile 设置为 SD 卡的首目录，其中调用 Environment 类定义的静态函数 getExternalStorageDirectory() 可以获得 SD 卡的首目录，Environment 类很强大，可以获得有关文件系统的很多信息，其常用的方法如图 10-5 所示。

	Public Methods	
static File	getDataDirectory()	
	Return the user data directory.	
static File	getDownloadCacheDirectory()	
	Return the download/cache content directory.	
static File	getExternalStorageDirectory()	
	Return the primary external storage directory.	
static File	getExternalStoragePublicDirectory(String type)	
	Get a top-level public external storage directory for placing files of a particular type.	
static String	getExternalStorageState(File path)	
	Returns the current state of the storage device that provides the given path.	
static String	getExternalStorageState()	
	Returns the current state of the primary "external" storage device.	
static File	getRootDirectory()	
	Return root of the "system" partition holding the core Android OS.	

图 10-5　Environment 类常用的方法

然后调用 load() 函数加载当前的目录信息，load() 函数的定义如下：

```
private void load() {
    File[] fl = cFile.listFiles();
    if (fl == null)
        return;

    cList.clear();

    int count = fl.length;
```

```
    for(int i=0; i<count; i++) {
        FileDesc fd = new FileDesc();
        fd.file = fl[i];
        fd.name = fl[i].getName();
        fd.length = fl[i].length();
        fd.type = fl[i].isDirectory()==true?"dir":"file";

        cList.add(fd);
    }

    this.notifyDataSetChanged();
}
```

这个函数首先获得当前目录的文件列表信息,并逐项处理,将它们存入 cList 结构对象中,然后通知 ListView 显示列表信息。

FileListAdapter 类的其他方法包括 getCount()、getItem()、getItemId(),我们不再进行解释,现在来看一下 getView()方法:

```
public View getView(int position, View convertView, ViewGroup parent) {
    FileDesc fd = cList.get(position);
    RelativeLayout rl;

    if (convertView != null)
        rl = (RelativeLayout)convertView;
    else
        rl = (RelativeLayout)inflater.inflate(R.layout.list_item, parent, false);

    rl.setTag(fd.file.getAbsolutePath());

    ImageView iv_type = (ImageView)rl.findViewById(R.id.id_li_image_view_type);
    if (fd.type.equalsIgnoreCase("dir"))
        iv_type.setImageResource(R.drawable.png1674);
    else
        iv_type.setImageResource(R.drawable.png0088);

    TextView tv_name = (TextView)rl.findViewById(R.id.id_li_text_view_name);
    tv_name.setText(fd.name);

    TextView tv_length = (TextView)rl.findViewById(R.id.id_li_text_view_length);
    tv_length.setText(fd.length + "");

    ImageView iv_play = (ImageView)rl.findViewById(R.id.id_li_image_view_play);
    iv_play.setImageResource(R.drawable.png1702);
    if ((fd.file.getAbsolutePath()).equalsIgnoreCase(currentPlaying))
        iv_play.setVisibility(View.VISIBLE);
    else
        iv_play.setVisibility(View.INVISIBLE);

    return rl;
}
```

在 getView()方法的前半部分,我们重用或新建一个列表项 RelativeLayout 对象。注意其中的

```
rl.setTag(fd.file.getAbsolutePath());
```

通过这条语句，我们为每个列表项设置一个 Tag 标志，进而在后续的操作中，我们使用 findViewWithTag()方法来获得需要的列表项。然后我们设置列表项应该显示的信息，包括文件类型、图标、文件名、文件长度，如果当前正在播放某个文件，则还要在该文件上显示播放图标。

FileListAdapter 类中的 goUP()方法、goDown()方法、getType()方法、getFile()方法及 setAndPlay()方法，将被 MainActivity 类调用，分别用于显示上一级目录、显示下一级目录、得到指定位置的列表项所显示的文件类型（目录还是普通文件）、得到指定位置的列表项的文件 File 对象及设置播放图标。现在对 setAndPlay()方法进行简单的解释，setAndPlay()方法的定义如下：

```java
public File setAndPlay(ListView lv, int position) {
    FileDesc fd = cList.get(position);
    RelativeLayout rl;

    if (currentPlaying != null) {
        rl = (RelativeLayout)lv.findViewWithTag(currentPlaying);
        if (rl != null) {
            ImageView iv_play =
                    (ImageView)rl.findViewById(R.id.id_li_image_view_play);
            iv_play.setVisibility(View.INVISIBLE);
        }
    }

    currentPlaying = fd.file.getAbsolutePath();
    rl = (RelativeLayout)lv.findViewWithTag(currentPlaying);
    if (rl != null) {
        ImageView iv_play = (ImageView)rl.findViewById(R.id.id_li_image_view_play);
        iv_play.setVisibility(View.VISIBLE);
    }

    return cList.get(position).file;
}
```

setAndPlay()函数的主要作用是使正在被播放的文件显示一个播放图标。其实现方式为：首先判断当前是否存在正在播放的文件，若有，则隐藏当前列表项的播放图标，并在即将播放的文件的列表项中显示播放图标。

最后，修改 MainActivity.java 文件，修改后的文件内容如下：

```java
package com.ttt.ex10multimedia02;

import java.io.File;
import java.io.IOException;

import com.ttt.ex10multimedia02.adapters.FileListAdapter;

import android.support.v7.app.ActionBar;
import android.support.v7.app.AppCompatActivity;
import android.media.AudioManager;
import android.media.MediaPlayer;
import android.media.MediaPlayer.OnErrorListener;
```

```java
import android.media.MediaPlayer.OnPreparedListener;
import android.os.Bundle;
import android.view.View;
import android.view.View.OnClickListener;
import android.widget.AdapterView;
import android.widget.AdapterView.OnItemClickListener;
import android.widget.ImageButton;
import android.widget.ListView;
import android.widget.TextView;
import android.widget.Toast;

public class MainActivity extends AppCompatActivity implements OnItemClickListener,
OnClickListener{
    private ListView lv;
    private ImageButton ib;
    private TextView tv;

    private FileListAdapter fla;

    private MediaPlayer mediaPlayer;

    @Override
    protected void onCreate(Bundle savedInstanceState) {
        super.onCreate(savedInstanceState);
        setContentView(R.layout.activity_main);

        ActionBar ab = this.getSupportActionBar();
        if (ab != null) ab.hide();

        lv = (ListView)this.findViewById(R.id.id_list_view);
        ib = (ImageButton)this.findViewById(R.id.id_image_button);
        tv = (TextView)this.findViewById(R.id.id_text_view_2);

        fla = new FileListAdapter(this);
        lv.setAdapter(fla);
        lv.setOnItemClickListener(this);

        ib.setOnClickListener(this);
    }

    @Override
    public void onItemClick(AdapterView<?> parent, View view, int position, long id) {
        String type = fla.getType(position);
        if (type.equalsIgnoreCase("dir")) {
            File ndir = fla.getFile(position);
            fla.goDown(ndir);
        }
        else {
            File nf = fla.setAndPlay(lv, position);
            tv.setText("正在播放：" + nf.getAbsolutePath());
            play(nf);
```

```java
        }
    }

    @Override
    public void onClick(View v) {
        int id = v.getId();
        if (id == R.id.id_image_button) {
            fla.goUp();
        }
    }

    private void play(File file) {
        if (mediaPlayer != null) {
            mediaPlayer.stop();
            mediaPlayer.release();
            mediaPlayer = null;
        }

        mediaPlayer = new MediaPlayer();
        mediaPlayer.setAudioStreamType(AudioManager.STREAM_MUSIC);
        mediaPlayer.setOnPreparedListener(new OnPreparedListener() {
            @Override
            public void onPrepared(MediaPlayer mp) {
                mediaPlayer.start();
            }
        });
        mediaPlayer.setOnErrorListener(new OnErrorListener() {
            @Override
            public boolean onError(MediaPlayer mp, int what, int extra) {
                Toast.makeText(MainActivity.this, "无法播放该文件",
                                        Toast.LENGTH_LONG).show();
                return false;
            }
        });

        try {
            mediaPlayer.setDataSource(file.getAbsolutePath());
            mediaPlayer.prepareAsync();
        } catch (IllegalArgumentException e) {
            e.printStackTrace();
        } catch (SecurityException e) {
            e.printStackTrace();
        } catch (IllegalStateException e) {
            e.printStackTrace();
        } catch (IOException e) {
            e.printStackTrace();
        }
    }
}
```

在这个程序的 onCreate()回调函数中，我们获得界面上各个组件的应用，并设置相应的监听接口。对列表项的点击监听是在 onItemClick()函数中完成的，内容如下：

```
public void onItemClick(AdapterView<?> parent, View view, int position, long id) {
    String type = fla.getType(position);
    if (type.equalsIgnoreCase("dir")) {
        File ndir = fla.getFile(position);
        fla.goDown(ndir);
    }
    else {
        File nf = fla.setAndPlay(lv, position);
        tv.setText("正在播放：" + nf.getAbsolutePath());
        play(nf);
    }
}
```

在这个函数中，程序首先判断所点击的列表项是目录还是普通文件，如果是目录，则调用 FileListAdapter 类的 goDown()函数在 ListView 中显示下一级目录的内容；如果是普通文件，则首先显示播放图标，在主界面的上方显示正在播放的文件的文件名，然后调用 play()函数播放所点击的文件。play()函数的定义如下：

```
private void play(File file) {
    if (mediaPlayer != null) {
        mediaPlayer.stop();
        mediaPlayer.release();
        mediaPlayer = null;
    }

    mediaPlayer = new MediaPlayer();
    mediaPlayer.setAudioStreamType(AudioManager.STREAM_MUSIC);
    mediaPlayer.setOnPreparedListener(new OnPreparedListener() {
        @Override
        public void onPrepared(MediaPlayer mp) {
            mediaPlayer.start();
        }
    });
    mediaPlayer.setOnErrorListener(new OnErrorListener() {
        @Override
        public boolean onError(MediaPlayer mp, int what, int extra) {
            Toast.makeText(MainActivity.this, "无法播放该文件",
                            Toast.LENGTH_LONG).show();
            return false;
        }
    });
    try {
        mediaPlayer.setDataSource(file.getAbsolutePath());
        mediaPlayer.prepareAsync();
    } catch (IllegalArgumentException e) {
        e.printStackTrace();
    } catch (SecurityException e) {
        e.printStackTrace();
    } catch (IllegalStateException e) {
        e.printStackTrace();
```

```
        } catch (IOException e) {
            e.printStackTrace();
        }
    }
```

首先判断当前是否正在播放某个文件，若是，则停止播放该文件。然后新建一个 MediaPlayer 对象，设置其可以播放音频，并设置 OnPreparedListener 接口。

现在的问题是：我们为什么要这么做呢？难道不能使用在 10.1.1 节中介绍的简单方法吗？回答是：不能。读者可以这样设想这个问题：假设我们要播放来自网络的某个音频，但网络的速率是不可预期的，因此，我们加载要播放的音频可能需要很长的时间，这将导致 Activity "响应超时"的问题。因此，对于大的或来自网络的音频，我们使用"异步加载"的方式来播放音频。这就是为什么要调用 setOnPreparedListener()方法的原因：使用异步加载。OnPreparedListener 接口只有一个方法：onPrepared()。它表示对音频的加载和解码已经完成，可以播放了。因此，在代码中，我们在 onPrepared()方法中直接播放音频。

对 MediaPlayer 对象，我们还同时设置了 OnErrorListener 接口来监听在加载、解码及播放过程中出现的任何错误：只是简单地显示一个错误消息而已。

在设置了相关接口后，我们调用如下代码：

```
            mediaPlayer.setDataSource(file.getAbsolutePath());
            mediaPlayer.prepareAsync();
```

来设置 MediaPlayer 即将播放的文件，并调用 prepareAsync()进行异步加载和解码。

完成后运行该程序，将显示如图 10-3 所示的界面。

在 Android Studio 中选择 Tools→Android → Android Device Monitor，即可打开 DDMS 来向模拟器中放置音频文件，如图 10-6 所示。

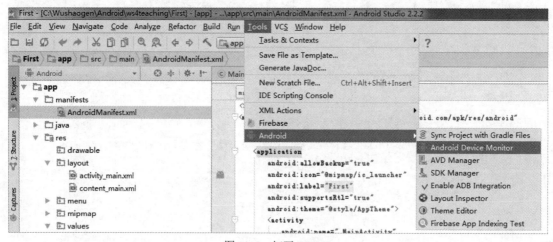

图 10-6　打开 DDMS

此时，将显示 DDMS 透视图，如图 10-7 所示。

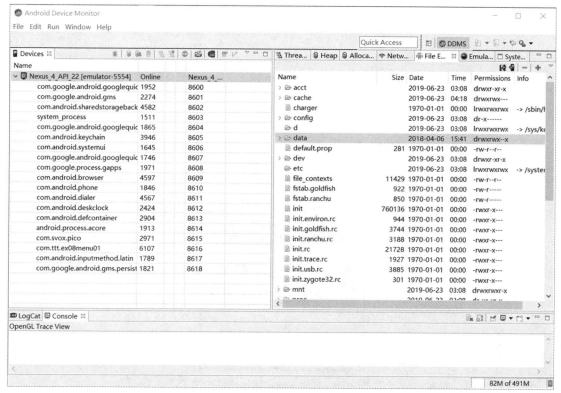

图 10-7　DDMS 透视图

选择"File Explorer"选项卡，通过 DDMS 透视图右上角的 工具即可操作右边的文件列表，包括将计算机中的文件下载到模拟器中、将模拟器中的文件加载到计算机中、删除模拟器中的文件或目录等操作。

10.2　同步练习一

将 10.1.2 节的例子加载到开发环境中，运行它，并进行如下修改：①使界面更加美观；②在列表中只显示能够播放的音频文件和子目录；③点击手机上的返回键图标显示当前目录的上一级目录的内容。

10.3　播放视频

我们可以使用 MediaPlayer 来播放视频，当然，要播放视频需要创建一个用于显示视频的组件。在此，我们不准备使用 Surface 组件，而是使用 Android 的便利组件 VideoView 来播放视频。下面通过一个简单的例子来介绍如何使用 VideoView 来播放视频。

在播放视频之前，首先使用 10.1.2 节中介绍的方法将一个 MP4 视频文件放置到模拟器的 SD 目录中，如图 10-8 所示。

图 10-8 示例视频文件 sample.mp4

完成后运行该程序,将播放该示例视频文件,如图 10-9 所示。

图 10-9 使用 VideoView 播放视频

我们来看一下这个程序的实现代码。在 Android Studio 中新建一个名为 Ex10MultiMedia03 的 Android 工程。首先修改 res/layout/activity_main.xml 文件,使之显示一个 VideoView 组件。修改后的 res/layout/activity_main.xml 文件内容如下:

```xml
<RelativeLayout xmlns:android="http://schemas.android.com/apk/res/android"
    android:layout_width="match_parent"
    android:layout_height="match_parent">

    <VideoView
        android:id="@+id/id_videoview"
        android:layout_width="match_parent"
        android:layout_height="match_parent"
        android:layout_centerInParent="true" />

</RelativeLayout>
```

然后修改 MainActivity.java 程序,修改后的程序如下:

```java
package com.ttt.ex10multimedia03;

import android.support.v7.app.AppCompatActivity;
import android.os.Bundle;
import android.os.Environment;
import android.widget.VideoView;

public class MainActivity extends AppCompatActivity {

    @Override
    protected void onCreate(Bundle savedInstanceState) {
```

```
        super.onCreate(savedInstanceState);
        setContentView(R.layout.activity_main);

        VideoView vv = (VideoView)this.findViewById(R.id.id_videoview);
        String file = Environment.getExternalStorageDirectory().getAbsolutePath() +
"/sample.mp4";
        System.out.println(file);
        vv.setVideoPath(file);
        vv.start();
    }
}
```

在 MainActivity 的 onCreate()回调函数中，首先从界面上获得 VideoView 的引用，然后，设置该 VideoView 要播放的视频文件，最后播放该文件。

10.4 同步练习二

修改 10.2 节同步练习一的程序，使之既可以播放音频也可以播放视频。提示：判断文件为视频后，在一个新的 Activity 中使用 VideoView 播放视频。

保存程序数据

在程序运行过程中，经常需要保存一些程序运行的数据，例如，为了避免用户每次登录时都被要求输入用户名和密码，当用户运行程序首次成功登录系统后，程序会将用户的登录数据保存下来，当用户下次再运行程序时，则不需要再次输入用户名、密码等数据。一般来说，程序都允许用户设置一些程序参数，例如，网络数据刷新时间、程序字体、每次显示的数据总数等，对于这些程序参数，一旦用户设置后，程序会将这些设置保存下来，当用户下次再运行程序时，将会自动采用这些已经设置好的参数。

我们可以使用 Android 提供的 SharedPreferences 来保存程序运行过程中的数据，也可以使用普通文件来保存这些数据，当然，我们还可以将这些数据保存在 SQLite 数据库中。

11.1 使用 SharedPreferences 保存程序数据

SharedPreferences 提供了一个基本框架，通过使用 SharedPreferences，可以非常方便地将程序运行数据保存下来。想要得到一个 SharedPreferences 对象，只需要在 Activity 中，调用 getSharedPreferences(String name, int mode)或 getPreferences(int mode)函数，其中的 name 参数是指定的用于保存数据的文件名，mode 参数设为 0 即可。getPreferences(int mode)函数只是简单地调用了 getSharedPreferences(String name, int mode)函数，并设定 name 参数为固定的值：调用这个函数的 Activity 的类名而已。

下面我们通过一个简单的例子来介绍如何使用 SharedPreferences 保存程序数据。程序首先显示一个登录界面，提示用户输入用户名和密码，一旦用户点击"登录"按钮，程序会将用户输入的用户名和密码保存下来，当用户再次运行这个程序时，程序会自动地将上次用户输入的用户名和密码填入到相应的输入框中。程序运行首界面如图 11-1 所示。

图 11-1 使用 SharedPreferences 的例子

现在来构建这个程序。新建一个名为 Ex11SaveData01 的 Android 工程。修改 res/layout/activity_main.xml 文件，修改后的文件内容如下：

```xml
<LinearLayout xmlns:android="http://schemas.android.com/apk/res/android"
    android:layout_width="match_parent"
    android:layout_height="match_parent"
    android:orientation="vertical">

    <EditText
        android:id="@+id/id_et_name"
        android:layout_width="match_parent"
        android:layout_height="wrap_content"
        android:hint="@string/text_name"
    />

    <EditText
        android:id="@+id/id_et_password"
        android:layout_width="match_parent"
        android:layout_height="wrap_content"
        android:inputType="textPassword"
        android:hint="@string/text_password"
    />

    <LinearLayout
        android:layout_width="match_parent"
        android:layout_height="wrap_content"
        android:orientation="horizontal"
    >

        <Button
        android:id="@+id/id_btn_login"
            android:layout_width="0dp"
            android:layout_height="wrap_content"
            android:layout_weight="1"
            android:text="@string/text_login"
        />

        <Button
        android:id="@+id/id_btn_reset"
            android:layout_width="0dp"
            android:layout_height="wrap_content"
            android:layout_weight="1"
            android:text="@string/text_reset"
        />

    </LinearLayout>

</LinearLayout>
```

这个布局文件很简单，显示了两个输入框和两个按钮。接下来，还需要修改 res/values/strings.xml 文件，在其中定义布局文件中用到的字符串引用，内容如下：

```xml
<?xml version="1.0" encoding="utf-8"?>
<resources>
```

```xml
    <string name="app_name">Ex11SaveData01</string>
    <string name="hello_world">Hello world!</string>

    <string name="text_name">用户名</string>
    <string name="text_password">密码</string>
    <string name="text_login">登录</string>
    <string name="text_reset">重置</string>

</resources>
```

现在修改 MainActivity.java 文件，使之显示主界面，并处理对按钮的点击事件：一旦用户点击"登录"按钮，则将用户输入的用户名和密码通过 SharedPreferences 保存下来。修改后的文件内容如下：

```java
package com.ttt.ex11savedata01;

import android.support.v7.app.AppCompatActivity;
import android.content.SharedPreferences;
import android.content.SharedPreferences.Editor;
import android.os.Bundle;
import android.view.View;
import android.view.View.OnClickListener;
import android.widget.Button;
import android.widget.EditText;

public class MainActivity extends AppCompatActivity implements OnClickListener{
    private Button btn_login, btn_reset;
    private EditText et_name, et_password;

    @Override
    protected void onCreate(Bundle savedInstanceState) {
        super.onCreate(savedInstanceState);
        setContentView(R.layout.activity_main);

        btn_login = (Button)this.findViewById(R.id.id_btn_login);
        btn_login.setOnClickListener(this);
        btn_reset = (Button)this.findViewById(R.id.id_btn_reset);
        btn_reset.setOnClickListener(this);

        et_name = (EditText)this.findViewById(R.id.id_et_name);
        et_password = (EditText)this.findViewById(R.id.id_et_password);

        SharedPreferences sp = this.getSharedPreferences("mimi",0);
        String name = sp.getString("name", "");
        et_name.setText(name);
        String password = sp.getString("password", "");
        et_password.setText(password);
    }

    @Override
    public void onClick(View v) {
        int id = v.getId();
```

```
        if (id == R.id.id_btn_login) {
            SharedPreferences sp = this.getSharedPreferences("mimi",0);
            Editor editor = sp.edit();
            editor.putString("name", et_name.getText().toString());
            editor.putString("password", et_password.getText().toString());
            editor.commit();
        }
        else {
            et_name.setText("");
            et_password.setText("");
        }
    }
}
```

在这个类的 onCreate()回调函数中，程序首先显示主界面，并获得界面上两个 EditText 组件及 Button 组件的应用，并设置按钮的点击监听接口，然后，通过如下代码：

```
            SharedPreferences sp = this.getSharedPreferences("mimi",0);
```

获得一个 SharedPreferences 对象。注意 getSharedPreferences("mimi", 0)语句的功能：它将在应用程序所在的安装目录的特定子目录下，也就是在/data/data/com.ttt.ex11savedata01/shared_prefs 目录下，检查是否存在名称为"mimi.xml"的文件，如果存在，则基于这个文件中已有的内容创建一个 SharedPreferences 对象，然后，再通过如下语句：

```
            String name = sp.getString("name", "");
            et_name.setText(name);
            String password = sp.getString("password", "");
            et_password.setText(password);
```

从这个 SharedPreferences 对象获得相应的指定的 key 键所对应的 value 值。在程序中，就是分别获得 key 为 "name" 和 key 为 "password" 的键所对应的值。如果这是第一次运行这个程序，显然，我们不能获得相应的值，因此，getString("name", "")和 getString("password","")都将返回第二个参数所给出的空串。

现在我们再来看看 onClick()事件处理函数，当用户点击"登录"按钮时，我们通过如下语句：

```
            SharedPreferences sp = this.getSharedPreferences("mimi",0);
            Editor editor = sp.edit();
            editor.putString("name", et_name.getText().toString());
            editor.putString("password", et_password.getText().toString());
            editor.commit();
```

获得一个 SharedPreferences 对象，然后调用这个对象的 edit()方法，也就是告诉 SharedPreferences 对象程序要向这个 SharedPreferences 对象写出 "key/value" 对，进而通过得到的 Editor 对象，调用它的 putString()方法，将用户在输入框输入的用户名和密码以 key 为 "name" 和 "password" 将相应的值写入到 SharedPreferences 对象中。

现在运行这个程序，在第一次显示界面后（见图 11-1），在两个输入框输入信息，然后点击"登录"按钮，再点击 Android 的退出键退出程序。之后，再次运行这个程序，将在相应的输入框中显示上次用户输入的信息。

在本质上，Android 的 SharedPreferences 其实就是在程序安装目录下的名称为 shared_prefs 的子目录下创建了以 getSharedPreferences 的第一个参数为文件名的特定的 XML 文件。通过 DDMS，选择 "File Explorer" 选项卡，可以清晰地看到程序所创建的 XML 文件，如图 11-2

所示。

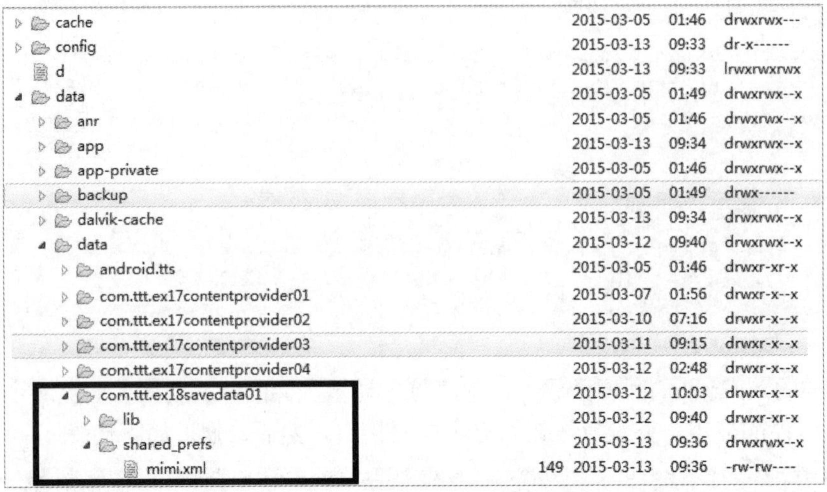

图 11-2　SharedPreferences 所对应的文件

我们还可以使用 DDMS 的"File Explorer"选项卡的右上角的工具 ![icon] 将 mimi.xml 文件从手机模拟器上复制到计算机中来。打开得到的文件，显示如下内容：

```
<?xml version='1.0' encoding='utf-8' standalone='yes' ?>
<map>
<string name="password">12345</string>
<string name="name">bill gates</string>
</map>
```

其中的 bill gates 和 12345 是我们在主界面的输入框中输入的内容。

11.2　同步练习一

完善 11.1 节的例子，做出如下修改：当用户点击"登录"按钮时，不仅在 SharedPreferences 中保存用户输入的用户名和密码，同时还保存用户登录的日期和时间。当用户再次运行程序时，不仅将上次用户输入的用户名和密码自动填入到相应输入框中，还通过一个 Toast 显示上次登录的日期和时间。完成后，运行程序，并使用 DDMS 视图观察应用程序的安装目录下文件的变化和文件结构。

11.3　设置程序首选项

Android 应用程序通常会包括一些设置以便用户可以改变程序的运行特征。例如，设置程序的界面风格、网络的刷新时间间隔或其他的一些与应用程序本身相关的运行参数。为了使 Android 应用程序与 Android 自带的"系统设置"应用具有一致的外观，Android 提供了一个称为 Preference 的 API。可以使用这个 API 来构建应用程序的首选项设置。

如图 11-3 所示就是一个典型的应用程序首选项设置界面。在这个界面中，包含各种可能的设置，例如：复选框选项表示是否选中某个特征，具体值参数表示可以设置某个参数的值等。

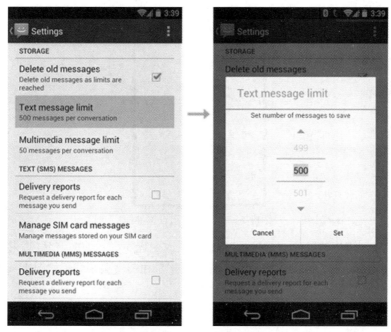

图 11-3　典型的应用程序首选项设置界面

Preference API 提供了实现如图 11-3 所示的程序首选项设置界面所需的所有组件,这些首选项组件所设置的值都以"key/value"的方式保存在应用程序安装目录下的 shared_prefs 子目录下文件名为"程序包名_preferencex.xml"的 XML 文件中,可以在首选项中保存的值包括 Boolean、Float、Int、Long、String 及 String 数组。常用的 Preference 的组件包括如下内容。

(1) CheckBoxPreference。

显示一个复选框用于设置某个特性是否可运行。

(2) ListPreference。

显示一个列表框,并在列表框中显示一组单选按钮,用于选择某个特征。

(3) EditTextPreference。

显示一个输入框,用于输入某个值。

下面通过一个例子来说明如何设置和使用应用程序首选项:我们通过使用应用程序首选项来设置显示文字"Hello World!"的样式,可设置的首选项包括是否以动画形式显示界面文字、文字的颜色、文字大小、文字的语种(中文或英文),运行效果如图 11-4 所示。

图 11-4　Ex11SaveData02 程序运行首界面

首先选择 Android 模拟器上的菜单,然后选择"Settings"选项,程序首选项设置界面如图 11-5 所示。

图 11-5　程序首选项设置界面

此时，可以设置程序的各个首选项，例如，如果改变程序显示的文字颜色，将其改为黑色，并点击返回按钮回到主界面，则程序主界面上的文字颜色将变为黑色，如图 11-6 所示。当然，还可以改变文字的语种等。

图 11-6　文字的颜色改为黑色

现在来构建这个程序，新建一个名为 Ex11SaveData02 的 Android 工程。修改 res/layout/activity_main.xml 文件，使其显示一个 TextView。修改后的文件内容如下：

```xml
<LinearLayout xmlns:android="http://schemas.android.com/apk/res/android"
    android:layout_width="match_parent"
    android:layout_height="match_parent">

    <TextView
        android:id="@+id/id_textview"
        android:layout_width="match_parent"
        android:layout_height="match_parent"
        android:gravity="center"
        android:text="@string/text_hello_world_en" />

</LinearLayout>
```

为了能够显示程序首选项设置界面，我们需要在 res 工程目录下，新建一个 XML 子目录，并在该目录下新建一个名为 my_settings.xml 文件（文件名可以是任意合法的文件名），修改后该文件的内容如下：

```xml
<?xml version="1.0" encoding="utf-8"?>
<PreferenceScreen xmlns:android="http://schemas.android.com/apk/res/android">
    <CheckBoxPreference
        android:key="pref_animation"
        android:title="@string/text_pref_animation"
        android:summary="@string/text_pref_animation_summ"
        android:defaultValue="false" />
    <ListPreference
        android:key="pref_lang_type"
        android:title="@string/text_pref_lang_type"
        android:dialogTitle="@string/text_pref_lang_type"
        android:entries="@array/pref_lang_type_entries"
        android:entryValues="@array/pref_lang_type_values"
        android:defaultValue="1" />
    <PreferenceCategory
        android:title="@string/text_pref_appearence_title">
        <EditTextPreference
           android:key="pref_text_size"
           android:title="@string/text_pref_text_size"
           android:summary="@string/text_pref_text_size_summ"
           android:defaultValue="16"/>
        <EditTextPreference
           android:key="pref_text_color"
           android:title="@string/text_pref_text_color"
           android:summary="@string/text_pref_text_color_summ"
           android:defaultValue="#FFFFFFFF"/>
    </PreferenceCategory>
</PreferenceScreen>
```

在这个首选项文件中，我们使用<PreferenceScreen>标签来表示要构建一个首选项文件，然后在这个标签下，我们通过如下语句：

```xml
<CheckBoxPreference
    android:key="pref_animation"
    android:title="@string/text_pref_animation"
    android:summary="@string/text_pref_animation_summ"
    android:defaultValue="false" />
```

来构建一个设置是否显示动画的选项，其中的

```
android:key="pref_animation"
```

和

```
android:defaultValue="false" />
```

用来设定该选项的 key 和其默认值，也就是"key/value"对。并通过 title 属性和 summary 属性来设定在首选项界面中显示的提示信息。类似地，我们通过如下语句：

```xml
<ListPreference
    android:key="pref_lang_type"
    android:title="@string/text_pref_lang_type"
    android:dialogTitle="@string/text_pref_lang_type"
    android:entries="@array/pref_lang_type_entries"
    android:entryValues="@array/pref_lang_type_values"
    android:defaultValue="1" />
```

来设定一个选择文字显示语种的列表选项，其中的

```
        android:key="pref_lang_type"
```

和

```
        android:defaultValue="1" />
```

用来设定"key/value"对，使用语句：

```
        android:title="@string/text_pref_lang_type"
        android:dialogTitle="@string/text_pref_lang_type"
        android:entries="@array/pref_lang_type_entries"
        android:entryValues="@array/pref_lang_type_values"
```

分别设置在首选项界面中显示的提示信息、打开列表框时显示的提示信息、列表框显示的列表信息及其对应的选项值。由于文字的大小和颜色都是设置文字外观的，因此，我们使用语句：

```
    <PreferenceCategory
```

对这两项设置进行分组：会将它们显示在一个分组中。其中的

```
    <PreferenceCategory
        android:title="@string/text_pref_appearence_title">
```

用于设置分组和分组显示的名称。然后，在这一分组中，我们设置两个输入框用于设定文字的大小和颜色，其方法与设置动画的属性类似。

在布局文件和这个首选项配置文件中，用到了一些字符串引用，因此，需要修改 res/values/strings.xml 文件，在其中定义需要的引用，修改后的文件内容如下：

```
<?xml version="1.0" encoding="utf-8"?>
<resources>

    <string name="app_name">Ex11Savedata02</string>
    <string name="action_settings">Settings</string>

    <string name="text_hello_world_en">Hello world!</string>
    <string name="text_hello_world_cn">你好，世界！</string>

    <string name="text_pref_animation">以动画形式显示文字</string>
    <string name="text_pref_animation_summ">选中该选项，
                                可以以动画的形式显示界面文字</string>
    <string name="text_pref_lang_type">选择显示文字的语种</string>

    <string-array name="pref_lang_type_entries">
        <item>English</item>
        <item>中文</item>
    </string-array>

    <string-array name="pref_lang_type_values">
        <item>1</item>
        <item>2</item>
    </string-array>

    <string name="text_pref_appearence_title">设置文字显示的外观</string>
    <string name="text_pref_text_size">文字字体大小</string>
    <string name="text_pref_text_size_summ">文字字体大小，请输入一个数值</string>
    <string name="text_pref_text_color">文字字体颜色</string>
```

```xml
    <string name="text_pref_text_color_summ">文字字体颜色,请输入一个数值,
                                            格式:#FFFFFFFF</string>
</resources>
```

为了显示动画,我们还需要在 res 目录下新建一个 anim 的子目录,并在其目录下新建一个名为 my_scale.xml 的动画文件,修改该文件后的内容如下:

```xml
<?xml version="1.0" encoding="utf-8"?>
<scale xmlns:android="http://schemas.android.com/apk/res/android"
    android:interpolator="@android:anim/accelerate_interpolator"
    android:fromXScale="0"
    android:toXScale="1"
    android:fromYScale="0"
    android:toYScale="1"
    android:duration="4000"
    android:pivotX="50%"
    android:pivotY="50%"
    android:fillAfter="true">
</scale>
```

现在构建首选项 Activity,在 src 目录下的 com.ttt.ex11savedata02 包下,新建一个名为 MySettingsActivity 的 Java 类文件,修改该文件为如下内容:

```java
package com.ttt.ex11savedata02;

import android.os.Bundle;
import android.preference.PreferenceActivity;

public class MySettingsActivity extends PreferenceActivity {
    public static String PREF_ANIMATION = "pref_animation";
    public static String PREF_LANG_TYPE = "pref_lang_type";
    public static String PREF_TEXT_SIZE = "pref_text_size";
    public static String PREF_TEXT_COLOR = "pref_text_color";

    @SuppressWarnings("deprecation")
    @Override
    public void onCreate(Bundle savedInstanceState) {
        super.onCreate(savedInstanceState);
        addPreferencesFromResource(R.xml.my_settings);
    }
}
```

在这个首选项 Activity 的变量定义中,我们定义了几个与 my_settings.xml 首选项文件中同名的字符串常量,它们用于设置选项的 key:

```java
public static String PREF_ANIMATION = "pref_animation";
public static String PREF_LANG_TYPE = "pref_lang_type";
public static String PREF_TEXT_SIZE = "pref_text_size";
public static String PREF_TEXT_COLOR = "pref_text_color";
```

通过这几个 key,我们就可以获得在首选项中设置的值。然后,再看一下这个类的 onCreate() 回调函数,代码如下:

```java
public void onCreate(Bundle savedInstanceState) {
    super.onCreate(savedInstanceState);
    addPreferencesFromResource(R.xml.my_settings);
```

```
        }
```

通过如下函数:

```
        addPreferencesFromResource(R.xml.my_settings);
```

就可以显示首选项界面，并且用户可以修改首选项的值。

现在回到MainActivity类，看一下如何获得和使用在首选项Activity中所设置的首选项值。修改 MainActivity.java 文件为如下内容：

```
package com.ttt.ex11savedata02;

import android.support.v7.app.AppCompatActivity;
import android.content.Intent;
import android.content.SharedPreferences;
import android.content.SharedPreferences.OnSharedPreferenceChangeListener;
import android.graphics.Color;
import android.os.Bundle;
import android.preference.PreferenceManager;
import android.view.animation.Animation;
import android.view.animation.AnimationUtils;
import android.widget.TextView;

public class MainActivity extends AppCompatActivity implements OnSharedPreferenceChangeListener{
    private TextView tv;

    @Override
    protected void onCreate(Bundle savedInstanceState) {
        super.onCreate(savedInstanceState);
        setContentView(R.layout.activity_main);

        tv = (TextView) findViewById(R.id.id_textview);

        SharedPreferences settings = PreferenceManager.getDefaultSharedPreferences(this);
        settings.registerOnSharedPreferenceChangeListener(this);

        String lang_type = settings.getString(MySettingsActivity.PREF_LANG_TYPE, "1");
        if (lang_type.equalsIgnoreCase("1"))
            tv.setText(R.string.text_hello_world_en);
        else
            tv.setText(R.string.text_hello_world_cn);

        String text_size = settings.getString(MySettingsActivity.PREF_TEXT_SIZE, "16");
        tv.setTextSize(Float.valueOf(text_size));

        String text_color = settings.getString(MySettingsActivity.PREF_TEXT_COLOR,
                                                                "#FFFFFFFF");
        tv.setTextColor(Color.parseColor(text_color));

        Boolean animation = settings.getBoolean(MySettingsActivity.PREF_ANIMATION,
                                                                        false);
        if (animation == true) {
            Animation scale = AnimationUtils.loadAnimation(this, R.anim.my_scale);
            tv.startAnimation(scale);
```

```
        }
    }

    @Override
    public void onSharedPreferenceChanged(SharedPreferences settings, String key) {
        if (key.equalsIgnoreCase(MySettingsActivity.PREF_ANIMATION)) {
            Boolean animation = settings.getBoolean(
                                    MySettingsActivity.PREF_ANIMATION, false);
            if (animation == true) {
                Animation scale = AnimationUtils.loadAnimation(this, R.anim.my_scale);
                tv.startAnimation(scale);
            }
        }
        else if (key.equalsIgnoreCase(MySettingsActivity.PREF_LANG_TYPE)) {
            String lang_type = settings.getString(
                                    MySettingsActivity.PREF_LANG_TYPE, "1");
            if (lang_type.equalsIgnoreCase("1"))
                tv.setText(R.string.text_hello_world_en);
            else
                tv.setText(R.string.text_hello_world_cn);
        }
        else if (key.equalsIgnoreCase(MySettingsActivity.PREF_TEXT_COLOR)) {
            String text_color = settings.getString(
                                    MySettingsActivity.PREF_TEXT_COLOR, "#FFFFFFFF");
            tv.setTextColor(Color.parseColor(text_color));
        }
        else if (key.equalsIgnoreCase(MySettingsActivity.PREF_TEXT_SIZE)) {
            String text_size = settings.getString(MyScttingsActivity.PREF_TEXT_SIZE, "16");
            tv.setTextSize(Float.valueOf(text_size));
        }
    }
}
```

在该类的 onCreate() 回调函数中，显示程序界面并获得 TextView 的引用，然后调用如下代码：

```
SharedPreferences settings = PreferenceManager.getDefaultSharedPreferences(this);
settings.registerOnSharedPreferenceChangeListener(this);
```

得到程序首选项对象，并设置首选项的值被改变时的监听函数：当首选项中任何一个值被改变时将调用该接口的 onSharedPreferenceChanged() 方法来实时改变文本框的显示特性。然后，我们通过如下代码从首选项对象中获得各个选项的设置参数值，并根据所选择的参数改变文本框的显示特性：

```
String lang_type = settings.getString(MySettingsActivity.PREF_LANG_TYPE, "1");
if (lang_type.equalsIgnoreCase("1"))
    tv.setText(R.string.text_hello_world_en);
else
    tv.setText(R.string.text_hello_world_cn);

String text_size = settings.getString(MySettingsActivity.PREF_TEXT_SIZE, "16");
tv.setTextSize(Float.valueOf(text_size));
```

```
        String text_color = settings.getString(MySettingsActivity.PREF_TEXT_COLOR,
                                                                "#FFFFFFFF");
        tv.setTextColor(Color.parseColor(text_color));

        Boolean animation = settings.getBoolean(MySettingsActivity.PREF_ANIMATION,
                                                                false);
        if (animation == true) {
            Animation scale = AnimationUtils.loadAnimation(this, R.anim.my_scale);
            tv.startAnimation(scale);
        }
    }
```

现在看一下 onSharedPreferenceChanged()函数,即当首选项中的任何一个设置参数值被改变时被调用的函数,这个函数根据被改变的参数来实时改变文本框的显示特性,内容如下:

```
@Override
public void onSharedPreferenceChanged(SharedPreferences settings, String key) {
    if (key.equalsIgnoreCase(MySettingsActivity.PREF_ANIMATION)) {
        Boolean animation = settings.getBoolean(
                            MySettingsActivity.PREF_ANIMATION, false);
        if (animation == true) {
            Animation scale = AnimationUtils.loadAnimation(this, R.anim.my_scale);
            tv.startAnimation(scale);
        }
    }
    else if (key.equalsIgnoreCase(MySettingsActivity.PREF_LANG_TYPE)) {
        String lang_type = settings.getString(
                            MySettingsActivity.PREF_LANG_TYPE, "1");
        if (lang_type.equalsIgnoreCase("1"))
            tv.setText(R.string.text_hello_world_en);
        else
            tv.setText(R.string.text_hello_world_cn);
    }
    else if (key.equalsIgnoreCase(MySettingsActivity.PREF_TEXT_COLOR)) {
        String text_color = settings.getString(
                    MySettingsActivity.PREF_TEXT_COLOR, "#FFFFFFFF");
        tv.setTextColor(Color.parseColor(text_color));
    }
    else if (key.equalsIgnoreCase(MySettingsActivity.PREF_TEXT_SIZE)) {
        String text_size = settings.getString(MySettingsActivity.PREF_TEXT_SIZE, "16");
        tv.setTextSize(Float.valueOf(text_size));
    }
}
```

我们通过选择菜单的"Settings"选项来打开 Activity 界面。现在运行这个程序,即可得到如图 11-5 所示的界面。

11.4 同步练习二

将 11.3 节中的例子复制到你的开发环境中并运行它,观察程序的运行效果,并做出如下修改:在首选项界面中,添加一项新的选项设置,用于设置动画的执行时间。提示:在获得动画对象后,使用 setDuration()函数来设置动画执行的时间;同时,对于动画执行时间这个选项,只有在执行动画时才有效,因此,还需要使用 android:dependency 首选项属性来配置这个首选项的依赖。

11.5 在程序目录下存储程序数据

任何一个 Android 应用程序,在被安装到 Android 系统中时,系统都将在/data/data 目录下,以这个应用程序的包名为名称为这个应用程序创建一个唯一的子目录,我们可以在这个子目录下创建只有自己程序才可以访问的子目录或文件。本节将介绍如何在应用程序的私有目录下创建和使用文件。

Activity 提供了如下的用于操作应用程序私有目录的方法(这些方法是在 Context 类中定义的,因为 Activity 是 Context 的子类,自然也继承了这些方法)。

(1) File Activity.getFilesDir()。

返回应用程序私有目录全路径的 File 对象。例如,如果应用程序的包名为 com.ttt.mysample,则这个 File 对象的全路径为/data/data/com.ttt.mysample/files。

(2) File getDir(String name,int mode)。

在应用程序的私有目录下创建或返回一个名字为 "app_" + name 参数的子目录 File 对象,设置 mode 参数为 0 即可。

(3) boolean deleteFile(String name)。

在应用程序的私有目录下删除一个名称为 name 的文件。

(4) String[] fileList()。

返回应用程序私有目录下的所有子目录或文件名。

(5) FileOutputStream openFileOutput(String name, int mode)。

在应用程序的私有目录下,打开或新建一个名称为 name 的文件用于写入数据,设置 mode 参数为 0 即可。

(6) FileInputStream openFileInput(String name)。

在应用程序的私有目录下,打开名称为 name 的文件用于从中读取数据。

除了上面这些方法,还可以使用 java.io 中提供的类及方法对应用程序私有目录下的任何文件或目录进行操作。因此,如果读者忘记了 Java 的流操作,则需要复习一下之前学习的 Java 语言。

11.6 同步练习三

参照 11.5 节中所介绍的各个文件访问方法的功能,编写一个简单的 Android 应用程序。提示:可以简单地输出相关信息即可。

11.7 访问外部存储器

对于配备了外部 SD 卡的 Android 设备,可以使用 Android 提供的一个工具类 Environment 来检查外部 SD 卡的状态及获得 SD 卡上的特定子目录,例如:是否插入 SD 卡、SD 卡当前是否可读、SD 卡当前是否可写、获得 SD 卡中特定子目录 File 对象等。

在读/写 SD 卡之前,需要为应用程序在 AndroidManifest.xml 文件中申请相应的权限。例如，为了读/写 SD 卡，需要编写如下代码：

```
<uses-permission android:name="android.permission.WRITE_EXTERNAL_STORAGE" />
```

而如果只需要读 SD 卡，则需要编写如下代码：

```
<uses-permission android:name="android.permission.READ_EXTERNAL_STORAGE" />
```

11.7.1 检查 SD 卡状态

在读/写外部 SD 卡之前，需要检查 SD 卡的状态，使用 Environment 的 getExternalStorageState()函数获得 SD 卡的状态，其状态包括 MEDIA_UNKNOWN、MEDIA_REMOVED、MEDIA_UNMOUNTED、MEDIA_CHECKING、MEDIA_NOFS、MEDIA_MOUNTED、MEDIA_MOUNTED_READ_ONLY、MEDIA_SHARED、MEDIA_BAD_REMOVAL、MEDIA_UNMOUNTABLE，它们的含义是不言自明的。例如，下面的代码片段用于检查外部 SD 卡当前的状态是否可读/写：

```java
/* 检查外部 SD 卡当前的状态是否可写，若是则返回 true，否则返回 false */
public boolean isExternalStorageWritable() {
    String state = Environment.getExternalStorageState();
    if (Environment.MEDIA_MOUNTED.equals(state)) {
        return true;
    }
    return false;
}

/* 检查外部 SD 卡当前的状态是否可读，若是则返回 true，否则返回 false */
public boolean isExternalStorageReadable() {
    String state = Environment.getExternalStorageState();
    if (Environment.MEDIA_MOUNTED.equals(state) ||
        Environment.MEDIA_MOUNTED_READ_ONLY.equals(state)) {
        return true;
    }
    return false;
}
```

11.7.2 获得 SD 卡上特定子目录的 File 对象

使用 Environment 的 getExternalStoragePublicDirectory(String type)方法可以获得 SD 卡上特定子目录的 File 对象。其中的 type 参数可取如下值：DIRECTORY_MUSIC、DIRECTORY_PODCASTS、DIRECTORY_RINGTONES、DIRECTORY_ALARMS、DIRECTORY_NOTIFICATIONS、DIRECTORY_PICTURES、DIRECTORY_MOVIES、DIRECTORY_DOWNLOADS、DIRECTORY_DCIM，这些参数的含义是不言自明的。例如，下面的这段代码片段表示将返回在 SD 卡上存放音乐文件的目录名：

```java
String fullpath = Environment.getExternalStoragePublicDirectory(
                  Environment.DIRECTORY_PICTURES).getAbsolutePath();
String path = Environment.getExternalStoragePublicDirectory(
              Environment.DIRECTORY_PICTURES).getName();
```

一旦得到了相关子目录，可以使用 java.io 中的相关类及其方法像操作普通文件一样操作 SD 卡上的文件或目录。

11.8 使用 SQLite 数据库保存程序数据

11.8.1 SQLite 数据库介绍

SQLite 是一个开源的、免费的数据库管理系统。与一般的基于 C/S 模式的数据库管理系统不同，如 Microsoft SQL Server、MySQL，这些基于 C/S 模式的数据库管理系统，数据库存储在称为服务器的计算机系统上，并通过数据库管理系统的服务器端程序来管理，需要使用数据库数据的程序，称为客户端程序，其通过某种通信协议，例如 TCP/IP 与数据库管理系统的服务器端程序通信来进行数据库数据的操作。SQLite 则全然不同，它不是基于 C/S 模式的，它只是一个 C 语言程序包（C 函数库），需要使用 SQLite 数据库的程序只需要调用这个程序包中的函数即可创建数据库、访问数据库中的数据等。

在 Android 系统中，Android 整合了 SQLite 数据库管理系统，将 SQLite 的 C 语言程序包进行了 Java 封装，提供了基于 Java 语言的类库。因此，在 Android 程序开发中，可以使用 SQLite 提供的 Java 接口来创建及访问 SQLite 数据库。

我们可以使用 SQL 规范中定义的 SQL 语言来创建和操作 SQLite 数据库。需要注意的是，对于所支持的数据类型，其他的基于 C/S 模式的数据库管理系统都提供了十分丰富的数据类型。例如 int、char、varchar、text、image、real 等。SQLite 也支持这些数据类型，但是，它采用自己的一套称为"Type Affinity"的机制进行数据类型自动映射，也就是将所有的数据类型都映射到如下的 5 种类型之一中。

（1）TEXT。字符串类型。
（2）NUMBERIC。精确表示的数值类型。
（3）INTEGER。整数类型。
（4）REAL 类型。采用 8 字节表示的 IEEE 浮点数据类型，与 NUMBERIC 不同，这个数据类型可能会有数据精度损失。
（5）BLOB 类型。二进制数据类型。

读者不需要对这套"Type Affinity"机制进行详细了解，在创建数据库表时，使用如上的 5 种数据类型就够了。

11.8.2 在 Android 中使用 SQLite 数据库

Android 提供了操作 SQLite 数据库的 Java 类库，对 SQLite 数据库完全支持，其中一个非常重要和常用的类就是 SQLiteOpenHelper 类，在需要创建和使用 SQLite 数据库时，应该继承这个类，并重写其中的 onCreate()方法来创建自己的数据表。

下面举一个例子来说明如何使用 SQLiteOpenHelper 类来创建自己的 SQLite 数据库，以及在数据库中创建自己需要的数据表。我们要创建的数据库名称为"Teach.db"，其中包含 3 张表：student 表、course 表和 score 表，其中分别存放学生的基本信息、课程的基本信息和学生成绩的基本信息。这 3 张表的结构定义如表 11-1、表 11-2 和表 11-3 所示。

表 11-1 student 表结构定义

字 段 名 称	数 据 类 型	备 注
_id	INTEGER	自增、主键
student_name	TEXT	学生姓名
student_birth	TEXT	学生出生日期
student_phone	TEXT	联系电话
student_photo	BLOB	学生头像

表 11-2 course 表结构定义

字 段 名 称	数 据 类 型	备 注
_id	INTEGER	自增、主键
course_name	TEXT	课程名称
course_memo	TEXT	课程介绍

表 11-3 score 表结构定义

字 段 名 称	数 据 类 型	备 注
_id	INTEGER	自增、主键
student_id	INTEGER	外键参照 student 表的 student_id 字段
course_id	INTEGER	外键参照 course 表的 course_id 字段
score_score	REAL	学生课程成绩

现在我们创建一个 Android 程序来创建需要的数据库和在数据库中创建需要的表。为此，新建一个名为 Ex11SQLite01 的 Android 工程，并在 src 中新建一个名为 com.ttt.ex11sqlite01.database 的包，在这个包下新建一个名为 MyDataBaseHelper 的 Java 类文件。现在修改 MyDataBaseHelper.java 文件，内容如下：

```java
package com.ttt.ex11sqlite01.database;

import android.content.Context;
import android.database.sqlite.SQLiteDatabase;
import android.database.sqlite.SQLiteOpenHelper;

public class MyDataBaseHelper extends SQLiteOpenHelper {

    private static final String DATABASE_NAME = "Teach.db";
    private static final int DATABASE_VERSION = 1;

    private static final String creat_student = "CREATE TABLE student ( " +
        "student_id INTEGER PRIMARY KEY," +
        "student_name TEXT, " +
        "student_birth TEXT, " +
        "student_phone TEXT, " +
        "student_photo BLOB " +
        ");";
    private static final String creat_course = "CREATE TABLE course ( " +
        "course_id INTEGER PRIMARY KEY, " +
        "course_name TEXT, " +
        "course_memo TEXT " +
        ");";
```

```java
    private static final String creat_score = "CREATE TABLE score ( " +
        "score_id INTEGER PRIMARY KEY," +
        "student_id INTEGER , " +
        "course_id INTEGER, " +
        "FOREIGN KEY(student_id) REFERENCES student(student_id) " +
        "FOREIGN KEY(course_id) REFERENCES course(course_id) " +
        ");";

    public MyDataBaseHelper(Context context) {
        super(context, DATABASE_NAME, null, DATABASE_VERSION);
    }

    @Override
    public void onCreate(SQLiteDatabase db) {
        db.execSQL(creat_student);
        db.execSQL(creat_course);
        db.execSQL(creat_score);
    }

    @Override
    public void onUpgrade(SQLiteDatabase db, int oldVersion, int newVersion) {
        db.execSQL("drop table score");
        db.execSQL("drop table student");
        db.execSQL("drop table course");

        onCreate(db);
    }
}
```

在这个类的变量定义中，我们定义了数据库的名称、数据库的版本和几个用于创建数据表的字符串常量。首先看一下其构造函数：

```java
public MyDataBaseHelper(Context context) {
    super(context, DATABASE_NAME, null, DATABASE_VERSION);
}
```

该构造函数直接调用了父类的构造函数：父类的构造函数会检查是否已经存在名为"Teach.db"且版本号为1的数据库。如果不存在，则会自动创建 Teach 数据库，并且调用 onCreate()函数创建需要的数据表，在这里，onCreate()函数创建了 3 张表：student 表、course 表和 score 表。如果已经存在名为"Teach.db"且版本号也为1的数据库，则不做任何操作；如果已经存在名为"Teach.db"的数据库，但是版本号小于由参数 DATABASE_VERSION 所指定的版本号，则会调用 onUpgrade()方法来升级数据库。在我们的 onUpgrade()函数中，只是简单地删除以前的表，然后重新创建所需要的表。

有了这个数据库操作辅助类，我们就可以轻松地操作数据库表的数据了。

现在我们继续完善这个程序：首先向 student 表中插入几条数据，然后通过一个 ListView 将 student 表中的数据显示出来。完成后运行效果如图 11-7 所示。

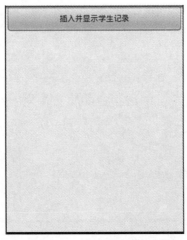

图 11-7　程序运行首界面

点击"插入并显示学生记录"按钮，将插入 3 条记录到 student 表中，然后从数据库的 student 表中将插入的数据读取出来并通过 ListView 显示出来，如图 11-8 所示。

图 11-8　插入 3 条记录并在 ListView 中显示

由于学生表中有学生头像，为了方便，我们直接将表示学生头像的 3 张图片放置在 res/drawable 工程目录下。现在修改 res/layout/activity_main.xml 主界面布局文件，修改后的内容如下：

```xml
<LinearLayout xmlns:android="http://schemas.android.com/apk/res/android"
    android:layout_width="match_parent"
    android:layout_height="match_parent"
    android:orientation="vertical">

    <Button
        android:id="@+id/id_button"
        android:layout_width="match_parent"
        android:layout_height="wrap_content"
        android:text="@string/text_button" />

    <ListView
        android:id="@+id/id_listview"
        android:layout_width="match_parent"
        android:layout_height="wrap_content"
```

```
    />

</LinearLayout>
```

这个布局文件很简单，只是显示了一个按钮和一个 ListView 而已。接着，在 res/layout 工程目录下新建一个列表项布局文件 list_item.xml，文件内容如下：

```xml
<?xml version="1.0" encoding="utf-8"?>
<LinearLayout xmlns:android="http://schemas.android.com/apk/res/android"
    android:layout_width="match_parent"
    android:layout_height="match_parent"
    android:orientation="horizontal" >

    <ImageView
        android:id="@+id/id_li_photo"
        android:layout_width="64dp"
        android:layout_height="64dp"
        android:scaleType="fitCenter"
        android:contentDescription="@string/hello_world"
    />

    <View
        android:layout_width="10dp"
        android:layout_height="64dp"
    />

    <LinearLayout
        android:layout_width="match_parent"
        android:layout_height="wrap_content"
        android:orientation="vertical"
    >

        <TextView
            android:id="@+id/id_li_id"
            android:layout_width="match_parent"
            android:layout_height="wrap_content"
        />

        <TextView
            android:id="@+id/id_li_name"
            android:layout_width="match_parent"
            android:layout_height="wrap_content"
        />

        <TextView
            android:id="@+id/id_li_birth"
            android:layout_width="match_parent"
            android:layout_height="wrap_content"
        />

        <TextView
            android:id="@+id/id_li_phone"
            android:layout_width="match_parent"
            android:layout_height="wrap_content"
```

```
        />

    </LinearLayout>

</LinearLayout>
```

在 ListView 的列表项中,显示了学生的头像、ID、姓名、出生日期和电话号码。我们还需要修改 res/values/strings.xml 文件,在其中定义按钮文字的字符串引用,修改后的 res/values/strings.xml 文件内容如下:

```xml
<?xml version="1.0" encoding="utf-8"?>
<resources>

    <string name="app_name">Ex11SQLite01</string>
    <string name="hello_world">Hello world!</string>
    <string name="action_settings">Settings</string>

    <string name="text_button">插入并显示学生记录</string>
</resources>
```

为了能在 ListView 中显示包含学生头像的数据记录,需要构建我们自己的 Adapter。为此,在 src 工程目录下,新建一个名为 com.ttt.ex11sqlite01.adapters 的包,并在该包下新建一个名为 MySimpleCursorAdapter 的 Java 类,修改其中的内容为如下代码:

```java
package com.ttt.ex11sqlite01.adapters;

import android.content.Context;
import android.database.Cursor;
import android.graphics.Bitmap;
import android.graphics.BitmapFactory;
import android.view.LayoutInflater;
import android.view.View;
import android.view.ViewGroup;
import android.widget.ImageView;
import android.widget.SimpleCursorAdapter;
import android.widget.TextView;

public class MySimpleCursorAdapter extends SimpleCursorAdapter {
    private int layout;
    private Cursor c;
    private String[] from;
    private int[] to;

    private LayoutInflater mInflater;

    @SuppressWarnings("deprecation")
    public MySimpleCursorAdapter(Context context, int layout, Cursor c,
            String[] from, int[] to) {
        super(context, layout, c, from, to);

        this.layout = layout;
        this.c = c;
        this.from = from;
```

```java
        this.to = to;

        mInflater = (LayoutInflater) context
                .getSystemService(Context.LAYOUT_INFLATER_SERVICE);
    }

    @Override
    public View getView(int position, View convertView, ViewGroup parent) {
        if (c.moveToNext()) {
            View v;
            v = mInflater.inflate(layout, parent, false);

            ImageView iv_photo = (ImageView) v.findViewById(to[0]);
            byte[] photo = c.getBlob(c.getColumnIndex(from[0]));
            Bitmap bm = BitmapFactory.decodeByteArray(photo, 0, photo.length);
            iv_photo.setImageBitmap(bm);

            TextView id = (TextView) v.findViewById(to[1]);
            String d = c.getString(c.getColumnIndex(from[1]));
            id.setText(d);

            TextView name = (TextView) v.findViewById(to[2]);
            d = c.getString(c.getColumnIndex(from[2]));
            name.setText(d);

            TextView birth = (TextView) v.findViewById(to[3]);
            d = c.getString(c.getColumnIndex(from[3]));
            birth.setText(d);

            TextView phone = (TextView) v.findViewById(to[4]);
            d = c.getString(c.getColumnIndex(from[4]));
            phone.setText(d);

            if (c.isLast()) c.moveToPosition(-1);

            return v;
        }

        return null;
    }
}
```

注意这个类的 getView() 方法。在这个方法中，我们从 Cursor 对象中获得学生头像的二进制数据，并通过 BitmapFactory 将代表头像的二进制数据组装成 Bitmap 对象，然后显示在 ImageView 组件中。对于其他的字符串数据，则直接显示在相应的 TextView 组件中。

现在修改 MainActivity.java 文件，使之显示主界面，监听对按钮的点击事件，并在对按钮的点击函数中插入学生记录到 student 表中，之后再从数据库的 student 表中读出数据并显示在 ListView 中，修改后的 MainActivity.java 文件的内容如下：

```java
package com.ttt.ex11sqlite01;

import java.io.ByteArrayOutputStream;
```

```java
import java.io.IOException;

import com.ttt.ex11sqlite01.adapters.MySimpleCursorAdapter;
import com.ttt.ex11sqlite01.database.MyDataBaseHelper;

import android.support.v7.app.AppCompatActivity;
import android.content.ContentValues;
import android.database.Cursor;
import android.database.sqlite.SQLiteDatabase;
import android.graphics.Bitmap;
import android.graphics.Bitmap.CompressFormat;
import android.graphics.BitmapFactory;
import android.os.Bundle;
import android.view.View;
import android.view.View.OnClickListener;
import android.widget.Button;
import android.widget.ListView;

public class MainActivity extends AppCompatActivity implements OnClickListener {
    private Button btn;
    private ListView lv;

    @Override
    protected void onCreate(Bundle savedInstanceState) {
        super.onCreate(savedInstanceState);
        setContentView(R.layout.activity_main);

        btn = (Button)this.findViewById(R.id.id_button);
        btn.setOnClickListener(this);

        lv = (ListView)this.findViewById(R.id.id_listview);
    }

    @Override
    public void onClick(View v) {
        MyDataBaseHelper helper = new MyDataBaseHelper(this);
        SQLiteDatabase dbw = helper.getWritableDatabase();

        ContentValues values = new ContentValues();
        ByteArrayOutputStream baos;
        Bitmap photo;

        //插入3条记录
        values.put("student_name", "张大卫");
        values.put("student_birth", "2011-01-01");
        values.put("student_phone", "13800138000");
        photo = BitmapFactory.decodeResource(this.getResources(), R.drawable.png0010);
        baos = new ByteArrayOutputStream();
        photo.compress(CompressFormat.PNG, 100, baos);
        values.put("student_photo", baos.toByteArray());
        try {
            baos.close();
        } catch (IOException e) {
```

```
            e.printStackTrace();
        }
        dbw.insert("student", null, values);

        values.put("student_name", "李丹丹");
        values.put("student_birth", "2001-11-09");
        values.put("student_phone", "13800138001");
        photo = BitmapFactory.decodeResource(this.getResources(), R.drawable.png0015);
        baos = new ByteArrayOutputStream();
        photo.compress(CompressFormat.PNG, 100, baos);
        values.put("student_photo", baos.toByteArray());
        try {
            baos.close();
        } catch (IOException e) {
            e.printStackTrace();
        }
        dbw.insert("student", null, values);

        values.put("student_name", "王芳");
        values.put("student_birth", "1990-10-10");
        values.put("student_phone", "13800138002");
        photo = BitmapFactory.decodeResource(this.getResources(), R.drawable.png1783);
        baos = new ByteArrayOutputStream();
        photo.compress(CompressFormat.PNG, 100, baos);
        values.put("student_photo", baos.toByteArray());
        try {
            baos.close();
        } catch (IOException e) {
            e.printStackTrace();
        }
        dbw.insert("student", null, values);

        dbw.close();

        SQLiteDatabase dbr = helper.getReadableDatabase();
        String[] columns = {"student_photo", "_id", "student_name",
                            "student_birth", "student_phone"};
        Cursor cur = dbr.query("student", columns, null, null, null, null, null);
        int[] li = {R.id.id_li_photo, R.id.id_li_id, R.id.id_li_name,
                    R.id.id_li_birth, R.id.id_li_phone};
        MySimpleCursorAdapter scad = new MySimpleCursorAdapter(MainActivity.this,
                                        R.layout.list_item, cur, columns, li);
        lv.setAdapter(scad);
    }
}
```

重点看一下按钮的处理函数 onClick()函数。这个函数首先创建一个我们之前定义的 MyDataBaseHelper 类的对象，并调用该对象的 getWritableDatabase()函数得到一个可以写入数据的数据库对象，内容如下：

```
MyDataBaseHelper helper = new MyDataBaseHelper(this);
SQLiteDatabase dbw = helper.getWritableDatabase();
```

然后，我们定义了几个数据变量，内容如下：

```
ContentValues values = new ContentValues();
ByteArrayOutputStream baos;
Bitmap photo;
```

其中的 ContentValues 是要写入数据表中的数据的字段名和数据值的"Key/Value"对。由于我们要插入头像到数据表中，因此，也定义了一个 Bitmap 变量和数据流变量。然后，我们可以使用这些变量向 student 表中插入一条数据记录，内容如下：

```
values.put("student_name", "张大卫");
values.put("student_birth", "2011-01-01");
values.put("student_phone", "13800138000");
photo = BitmapFactory.decodeResource(this.getResources(), R.drawable.png0010);
baos = new ByteArrayOutputStream();
photo.compress(CompressFormat.PNG, 100, baos);
values.put("student_photo", baos.toByteArray());
try {
   baos.close();
} catch (IOException e) {
   e.printStackTrace();
}
dbw.insert("student", null, values);
```

通过上面的代码，即可向 student 表中插入一条记录。类似地，我们还插入了其他两条记录。在插入数据记录后，通过如下代码：

```
SQLiteDatabase dbr = helper.getReadableDatabase();
String[] columns = {"student_photo", "_id", "student_name",
                    "student_birth", "student_phone"};
Cursor cur = dbr.query("student", columns, null, null, null, null, null);
int[] li = {R.id.id_li_photo, R.id.id_li_id, R.id.id_li_name,
            R.id.id_li_birth, R.id.id_li_phone};
MySimpleCursorAdapter scad = new MySimpleCursorAdapter(MainActivity.this,
                             R.layout.list_item, cur, columns, li);
lv.setAdapter(scad);
```

从数据库中将 student 表的记录全部读取出来，并通过 ListView 显示出来。

现在运行代码，即可显示如图 11-8 所示的界面。

现在有一个问题：我们所创建的 SQLite 数据库在手机的什么位置？首先需要知道的是，我们在手机上安装的应用程序，都会安装在手机的内部文件系统的/data/data 目录下。为了更加直观地看到这一点，打开 DDMS 视图界面，然后选择 DDMS 视图中的"File Explorer"选项卡，将显示如图 11-9 所示的手机文件系统列表。

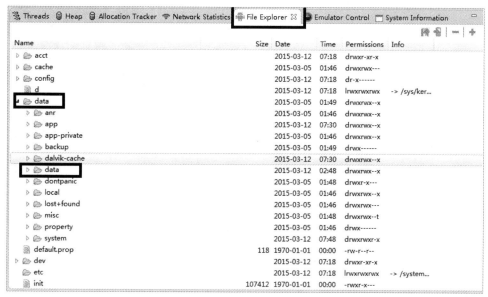

图 11-9 DDMS 视图下的手机文件系统列表

在如图 11-9 所示的手机文件目录中，展开/data/data 目录，就会看到我们已经安装的 com.ttt.ex11sqlite01 程序目录，展开这个目录，在 databases 目录下，就可以看到我们创建的名为"Teach.db"的数据库文件，还可以通过 DDMS 视图中的"File Explorer"右上角的工具按钮 ![] 将这个文件从手机上复制到计算机上。

每个 Android 应用程序，一旦安装到手机上，Android 系统会在手机文件系统的/data/data 目录下以应用程序的包名为名称为该应用程序创建一个应用程序目录，这个目录是该应用程序私有的，只有该应用程序和特权 root 用户才可以读/写这个目录及其子目录。

使用后台任务

在 Android 系统中，Activity 运行在称为 UI 线程的主线程中，并且系统对 Activity 的响应时间有严格的要求：也就是对每个用户操作的响应时长不能超过规定的时间长度，否则系统将出现异常，并且也会影响用户的使用体验。因此，在 Android 系统中，对于需要使用较长时间执行的功能都应该放在后台进行。不仅如此，Android 也要求对于需要使用较长时间或执行时间不确定的功能，例如网络通信等都必须放在后台执行。

在后台执行任务，完全可以使用 Java 的线程机制进行。在 Android 应用程序中，在后台执行程序任务是非常普遍的，Android SDK 也提供了相应的机制使在后台执行任务变得更加容易实现，包括使用 AsyncTask 类。本章将对如何在后台执行程序任务进行介绍。

12.1 使用 Java 线程执行后台任务

对于一些需要使用较长时间执行的程序任务，我们完全可以使用 Java 的线程机制，也就是 Thread 类来执行这些任务。下面我们通过一个简单的例子来看一下如何使用 Java 的 Thread 类来执行后台任务。

这个例子程序显示一个简单的时钟：在一个 TextView 中实时显示系统的当前日期和时间。程序运行效果如图 12-1 所示。

图 12-1　使用 Java 线程机制执行后台任务的例子首界面

此时，程序每隔 1 秒实时显示系统时间，点击"停止"按钮将停止实时显示，并且按钮上的文字将变为"启动"，再次点击该按钮，又将实时显示系统时间。

现在我们来构建这个例子程序，新建一个名为 Ex12Background01 的 Android 工程。修改 res/layout/activity_main.xml 文件，使之显示一个 TextView 和一个 Button，修改后的文件内容如下：

```xml
<RelativeLayout xmlns:android="http://schemas.android.com/apk/res/android"
    android:layout_width="match_parent"
    android:layout_height="match_parent">

    <TextView
        android:id="@+id/id_textview"
        android:layout_width="wrap_content"
        android:layout_height="wrap_content"
        android:layout_centerInParent="true"
        android:textSize="24sp" />

    <Button
        android:id="@+id/id_button"
        android:layout_width="match_parent"
        android:layout_height="wrap_content"
        android:layout_alignParentBottom="true"
        android:text="@string/text_button_stop"
        />

</RelativeLayout>
```

然后，还需要修改 res/values/strings.xml 文件，在其中定义在布局文件中用到的字符串引用，修改后的文件内容如下：

```xml
<?xml version="1.0" encoding="utf-8"?>
<resources>

    <string name="app_name">Ex12Background01</string>

    <string name="text_button_start">启动</string>
    <string name="text_button_stop">停止</string>

</resources>
```

最后，修改 MainActivity.java 文件，修改后的文件内容如下：

```java
package com.ttt.ex12background01;

import java.text.SimpleDateFormat;
import java.util.Date;
import java.util.Locale;

import android.support.v7.app.AppCompatActivity;
import android.os.Bundle;
import android.os.Handler;
import android.view.View;
import android.view.View.OnClickListener;
import android.widget.Button;
import android.widget.TextView;

public class MainActivity extends AppCompatActivity implements OnClickListener {
    private TextView tv;
    private Button btn;

    private boolean started;
```

```java
    private Handler handler;

    private Date d;
    private SimpleDateFormat sdf;

    @Override
    protected void onCreate(Bundle savedInstanceState) {
        super.onCreate(savedInstanceState);
        setContentView(R.layout.activity_main);

        started = true;
        d = new Date();

        tv = (TextView)this.findViewById(R.id.id_textview);
        sdf = new SimpleDateFormat("yyyy-MM-dd HH:mm:ss", Locale.CHINA);
        String ds = sdf.format(d);
        tv.setText(ds);

        btn = (Button)this.findViewById(R.id.id_button);
        btn.setOnClickListener(this);

        handler = new Handler();
        Thread t = new Thread(new MyTimer());
        t.start();
    }

    private class MyTimer implements Runnable {
        @Override
        public void run() {
            while(started) {
                try {
                    Thread.sleep(1000);
                } catch (InterruptedException e) {
                    e.printStackTrace();
                }

                handler.post(new Runnable(){
                    @Override
                    public void run() {
                        d.setTime(System.currentTimeMillis());
                        String ds = sdf.format(d);
                        tv.setText(ds);
                    }
                });
            }
        }
    }

    @Override
    public void onClick(View v) {
        Button b = (Button)v;

        if (started == true) {
```

```
            started = false;
            b.setText(R.string.text_button_start);
        }
        else {
            started = true;
            Thread t = new Thread(new MyTimer());
            t.start();
            b.setText(R.string.text_button_stop);
        }
    }
}
```

在类的变量定义中,我们定义了 started 变量,用于表示当前时钟是否正在运行,也就是是否在实时显示系统时间,同时,还定义了一个类型为 Handler 的变量 handler,这个变量的作用是什么呢?这需要从 Android 的 Activity 的工作机制说起。

Android 的每个 Activity 都运行在自己独立的线程中,这个线程通常称为 UI 主线程,通过这个 UI 主线程 Activity 实现了与用户的交互。为了保证界面交互的实时性和不影响用户体验,Android 规定 Activity 完成任何一次交互的时长都不能超过规定的时间(10 秒),若超过这个时间,Android 将会发生异常;同时,Android 还规定,Activity 界面上所显示的信息,只能由 UI 线程对其进行更新,其他任何线程都不能更新界面上所显示的信息,否则会发生异常。为了能够让其他非 UI 线程也能修改界面上的信息,Android 为每个 Activity 提供了一个默认的消息队列,需要修改界面信息的线程通过这个消息队列向 Activity 发送消息进而由 Activity 的主界面 UI 线程来修改界面信息,Handler 类就是为这个目的而设计的。

现在看一下这个类的 onCreate()回调函数。在这个函数中,首先显示主界面并获得界面上的 TextView 和 Button 的引用,在 TextView 组件上显示当前的日期和时间,然后设置按钮的点击响应函数,获得 Handler 对象用于非 UI 线程与 UI 主线程之间的通信。最后,启动一个后台线程,用于以 1 秒的时间间隔更新界面上的时间显示。后台线程的代码如下:

```
private class MyTimer implements Runnable {
    @Override
    public void run() {
        while(started) {
            try {
                Thread.sleep(1000);
            } catch (InterruptedException e) {
                e.printStackTrace();
            }

            handler.post(new Runnable(){
                @Override
                public void run() {
                    d.setTime(System.currentTimeMillis());
                    String ds = sdf.format(d);
                    tv.setText(ds);
                }
            });
        }
    }
}
```

在这个后台线程中，首先判断 started 是否为 true，若为 true，则表示要修改界面上的时间显示。为此，每隔 1 秒，通过 handler 对象向 UI 主线程发送一个消息：这个消息有点特殊，是要求主界面 UI 线程执行一个特定的函数，在这个特定的函数中，修改在界面上要显示的当前日期和时间。

对于按钮的点击处理函数则比较简单，只是启动或停止线程的执行而已：通过设置 started 变量的值为 false 来关闭后台线程的执行，当要再次启动后台线程时，重新创建一个后台线程即可。运行这个程序，即可得到如图 12-1 所示的结果。

12.2 同步练习一

将 12.1 节中的程序复制到你的开发环境中，并进行如下修改：每到一个整点，例如早上 8:00 或晚上 10:00，系统自动播放一段简短的音乐。提示：每到一个整点，启动一个后台线程来播放音乐。

12.3 使用 AsyncTask 执行后台任务

在 Android 程序中执行后台任务是一个普遍的要求，因此，Android 为了便于实现后台任务，提供了 AsyncTask 工具类。使用 AsyncTask，可以使应用程序在后台执行任务，并将任务的运行状态或结果显示在 UI 主界面线程中。

使用 AsyncTask，需要派生一个 AsyncTask 的子类，并且需要重写 AsyncTask 的 4 个方法，其方法如下所示。

（1）protected void onPreExecute()。

在执行后台任务之前需要执行的操作，例如进行一些初始化的操作，在这个操作中，可以直接将信息显示在 UI 主界面中。由于这个方法是在 UI 主线程中执行的，因此，这个方法要简短、高效。

（2）protected Result doInBackground(Params... params)。

在这个方法中实现需要在后台执行的任务，这个方法在一个新线程中执行，在通过 execute() 函数调用启动异步任务时传递的参数将传递给这个函数。在这个函数中，通过调用 publishProgress() 方法将任务执行过程中的状态信息传递给 UI 线程，同时，这个函数的返回值也将被传递给 UI 线程。

（3）protected void onProgressUpdate(Progress... values)。

在后台任务执行的过程中，后台任务可以将中间状态信息传递给这个函数，通过这个函数，可以将后台任务运行的中间状态传递给 UI 线程。

（4）protected void onPostExecute(Result result)。

当后台任务执行完毕后，可以使用这个函数将后台任务的运行结果通过 UI 线程显示在主界面中。

AsyncTask 类支持 3 个泛型的类，因此，在派生 AsyncTask 类的子类时，需要指出 3 个类型：第一个数据类型是传递给 doInBackground() 方法的 Params 的类型、第二个数据类型是传递给 onProgressUpdate() 方法的 Progress 的类型、第三个数据类型是传递给 onPostExecute() 方法的 Result 数据类型。对不需要使用的参数，我们都可以传递 Void 类型到 AsyncTask 泛型。

上面介绍了这么多基础内容，可能读者对如何使用 AsyncTask 还是比较模糊，下面举一个例子来说明如何使用 AsyncTask 来执行后台任务。使用 AsyncTask 实现 12.1 节中介绍的实时显示日期和时间的例子。运行效果如图 12-2 所示。

图 12-2　使用 AsyncTask 显示日期和时间

点击"停止"按钮，将显示如图 12-3 所示的界面。与 12.1 节中界面显示不同的是，我们使用 Toast 显示一个简单的信息框以告知用户当前时钟的状态。

图 12-3　停止实时显示日期和时间的界面

下面来构建这个程序。新建一个名为 Ex12Background02 的 Android 工程。修改 res/layout/activity_main.xml 文件，使之显示一个 TextView 组件和一个 Button 组件，修改后的文件内容如下：

```xml
<RelativeLayout xmlns:android="http://schemas.android.com/apk/res/android"
    android:layout_width="match_parent"
    android:layout_height="match_parent">

    <TextView
        android:id="@+id/id_textview"
        android:layout_width="wrap_content"
        android:layout_height="wrap_content"
        android:layout_centerInParent="true"
        android:textSize="24sp" />

    <Button
```

```xml
        android:id="@+id/id_button"
        android:layout_width="match_parent"
        android:layout_height="wrap_content"
        android:layout_alignParentBottom="true"
        android:text="@string/text_button_stop"
    />

</RelativeLayout>
```

然后，还需要修改 res/values/strings.xml 文件，在其中定义几个字符串引用，修改后的文件内容如下：

```xml
<?xml version="1.0" encoding="utf-8"?>
<resources>

    <string name="app_name">Ex12Background02</string>

    <string name="text_button_start">启动</string>
    <string name="text_button_stop">停止</string>

</resources>
```

最后，修改 MainActivity.java 文件，修改后的文件内容如下：

```java
package com.ttt.ex12background02;

import java.text.SimpleDateFormat;
import java.util.Date;
import java.util.Locale;
import java.util.concurrent.atomic.AtomicBoolean;

import android.support.v7.app.AppCompatActivity;
import android.os.AsyncTask;
import android.os.Bundle;
import android.view.View;
import android.view.View.OnClickListener;
import android.widget.Button;
import android.widget.TextView;
import android.widget.Toast;

public class MainActivity extends AppCompatActivity implements OnClickListener {
    private TextView tv;
    private Button btn;

    private AtomicBoolean started = new AtomicBoolean();
    private Date d;
    private SimpleDateFormat sdf;

    private MyAsyncTask mat;

    @Override
    protected void onCreate(Bundle savedInstanceState) {
        super.onCreate(savedInstanceState);
        setContentView(R.layout.activity_main);
```

```java
        started.set(false);
        d = new Date();

        tv = (TextView)this.findViewById(R.id.id_textview);
        sdf = new SimpleDateFormat("yyyy-MM-dd HH:mm:ss", Locale.CHINA);
        String ds = sdf.format(d);
        tv.setText(ds);

        btn = (Button)this.findViewById(R.id.id_button);
        btn.setOnClickListener(this);
    }

    @Override
    protected void onResume() {
        super.onResume();

        if (started.get() == false) {
            started.set(true);
            mat = new MyAsyncTask();
            mat.execute();
        }
    }

    @Override
    protected void onPause() {
        super.onPause();

        if (started.get() == true) {
            started.set(false);
        }
    }

    @Override
    public void onClick(View v) {
        if (started.get() == true) {
            started.set(false);
            btn.setText(R.string.text_button_start);
        }
        else {
            started.set(true);
            mat = new MyAsyncTask();
            mat.execute();
            btn.setText(R.string.text_button_stop);
        }
    }

    private class MyAsyncTask extends AsyncTask<Void, Void, Void> {
        @Override
        protected void onPreExecute() {
            Toast.makeText(MainActivity.this, "开始实时显示时间",
                                        Toast.LENGTH_SHORT).show();
        }
```

```
        @Override
        protected Void doInBackground(Void... params) {
            while(started.get() == true) {
                try {
                    Thread.sleep(1000);
                } catch (InterruptedException e) {
                    e.printStackTrace();
                    return null;
                }
                publishProgress();
            }
            return null;
        }

        @Override
        protected void onProgressUpdate (Void... values) {
            d.setTime(System.currentTimeMillis());
            String ds = sdf.format(d);
            tv.setText(ds);
        }

        @Override
        protected void onPostExecute (Void result) {
            Toast.makeText(MainActivity.this, "停止实时显示时间",
                                         Toast.LENGTH_SHORT).show();
        }
    }
}
```

在这个类的 onCreate()回调函数中，我们获得界面上组件的引用，并设置对按钮点击的响应处理接口。

我们重点看一下 MyAsyncTask 类的实现。在 MyAsyncTask 类中，由于 4 个需要重写的方法中均没有参数，因此，在 AsyncTask 的泛型中，我们使用了如下代码：

```
AsyncTask<Void, Void, Void>
```

来表示 MyAsyncTask 类的 3 个方法：doInBackground、onProgressUpdate 和 onPostExecute。这些方法都不需要参数（Void 就是"无"的意思）。现在看一下 onPreExecute()方法的实现，内容如下：

```
        @Override
        protected void onPreExecute() {
            Toast.makeText(MainActivity.this, "开始实时显示时间",
                                         Toast.LENGTH_SHORT).show();
        }
```

由于这个方法是在 UI 线程中执行的，因此，在这个方法中，我们可以放心地进行界面信息修改，在这里，我们使用 Toast 显示一个简短的信息。再看一下 doInBackground()方法的实现，内容如下：

```
        @Override
        protected Void doInBackground(Void... params) {
            while(started.get() == true) {
                try {
```

```
                Thread.sleep(1000);
            } catch (InterruptedException e) {
                e.printStackTrace();
                return null;
            }
            publishProgress();
        }
        return null;
    }
```

在这个函数中，当 started 为 true 时，程序每隔 1 秒调用一次 publishProgress()函数，进而调用 onProgressUpdate()函数来修改界面上显示的信息。注意，由于 onProgressUpdate()也是在 UI 线程中执行的，因此，我们也可以放心地修改界面上显示的信息，内容如下：

```
@Override
protected void onProgressUpdate (Void… values) {
    d.setTime(System.currentTimeMillis());
    String ds = sdf.format(d);
    tv.setText(ds);
}
```

最后，当在某个地方 started 被设置为 false 时，将导致 doInBackground()结束运行，进而系统将调用 onPostExecute()函数，在这里，我们再次使用 Toast 显示一个简短的信息，内容如下：

```
@Override
protected void onPostExecute (Void result) {
    Toast.makeText(MainActivity.this, "停止实时显示时间",
                                        Toast.LENGTH_SHORT).show();
}
```

编写完成 MyAsyncTask 类后，为了启动时钟线程的运行，我们在 Activity 的 onResume()回调函数中创建了 MyAsyncTask 类的对象，并调用其 execute()方法启动后台线程开始运行，进而修改界面上的时钟显示状态。

当我们结束应用程序时，需要停止后台线程的执行。因此，在 Activity 的 onPause()回调函数中，我们将 started 设置为 false 进而停止后台线程的运行以达到停止时钟显示的目的。

这里还需要说明的是，注意程序中的 started 变量，由于在 UI 线程及 MyAsyncTask 线程中均需要访问这个变量，为了保证该变量数据的完整性，我们使用了 java.util.concurrent.atomic.AtomicBoolean 类型的变量。通过将 started 定义为 AtomicBoolean 类型，我们可以使用 Java JDK 提供的并发控制机制来保证 started 变量在被多个线程使用时的数据完整性。类似地，对于在多线程中需要使用到的一个原始数据类型，都可以使用 java.util.concurrent.atomic 包中的类型来保证数据的完整性。

运行这个程序，即可得到如图 12-2 所示的结果。

12.4 使用 Service 完成后台任务

Service 是 Android 的组件之一，它没有 UI 接口，使用 Service 可以完成一些需要长时间在后台运行的任务。Android 的其他组件，如 Activity，可以启动一个 Service 运行。Service 一旦被启动，它将持续地在后台运行，直到被停止。Android 提供两种类型的 Service，分别为"启动式服务"和"绑定式服务",在本节,我们只对启动式服务进行介绍,对于绑定式服务及 IPC(进程间通信),读者可以参考 Android 的帮助文档。

与 Activity 一样,Service 也具有其固有的生命周期,我们通过重写这些生命周期回调函数来完成对 Service 的控制。Service 的生命周期如图 12-4 所示。

从如图 12-4 所示的 Service 的生命周期可以看出,当应用程序的组件调用 startService()方法时,将启动某个指定的 Service 运行,此时,如果该 Service 之前没有运行,系统将调用 Service 的 onCreate()回调函数初始化该 Service,之后将调用 Service 的 onStartCommand()方法;如果该 Service 之前已经被其他的某个应用程序组件启动,则将直接调用 Service 的 onStartCommand()方法,自此,Service 进入服务活动状态。当应用程序组件调用 stopService()方法停止 Service 运行或 Service 自己调用 stopSelf()方法停止运行时,系统将调用 Service 的 onDestroy()回调函数,自此,该 Service 的生命周期停止。

在此需要强调的是,虽然 Service 是后台服务,但是,Service 是运行在应用程序的主线程中的,也就是应用程序的 UI 线程中。Service 是后台服务,是指当启动 Service 的应用程序组件被停止后,Service 仍然在运行。例如,一个 Activity 启动了某个 Service 运行,但是,当 Service 被启动后,启动 Service 的 Activity 被用户关闭了。虽然 Activity 被关闭了,但是被 Activity 启动的 Service 仍然可以继续运行。这就是 Service 是后台服务的真正含义。

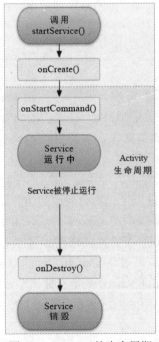

图 12-4 Service 的生命周期

因为 Service 运行在应用程序的主线程中,因此,在 Service 的各个回调函数中只能做一些简短的工作,对于需要长时间运行的任务则交给某个线程来运行。

下面我们举一个例子来说明 Service 的应用:使用 Service 在后台播放某个音乐。在第 10 章中我们介绍媒体播放程序时,是在 Activity 中播放音乐的,因此,当播放音乐的 Activity 退出时,系统将停止播放。现在我们做出如下修改:使用 Activity 来启动音乐的播放,当这个 Activity 退出时,系统将继续播放音乐。程序的运行效果如图 12-5 所示。

图 12-5 使用 Service 播放音乐的运行界面

可以点击"播放 1 号音乐"或"播放 2 号音乐"来播放音乐，此时，程序将启动一个 Service 来播放相应的音乐，同时，在系统的通知栏将显示一个播放图标，如图 12-6 所示。这时当退出界面 Activity 时，所播放的音乐将不会被停止。我们可以通过下拉通知栏再次启动界面 Activity 来对播放进行控制。

图 12-6 在系统通知栏显示一个通知

现在看一下如何构建这个程序。新建一个名为 Ex12Background03 的 Android 工程。修改 res/layout/activity_main.xml 布局文件，文件内容如下：

```xml
<RelativeLayout xmlns:android="http://schemas.android.com/apk/res/android"
    android:layout_width="match_parent"
    android:layout_height="match_parent">

    <Button
        android:id="@+id/id_button_1"
        android:layout_width="match_parent"
        android:layout_height="wrap_content"
```

```xml
        android:layout_alignParentTop="true"
        android:text="@string/text_button_1" />

    <Button
        android:id="@+id/id_button_2"
        android:layout_width="match_parent"
        android:layout_height="wrap_content"
        android:layout_below="@id/id_button_1"
        android:text="@string/text_button_2" />

    <Button
        android:id="@+id/id_button_3"
        android:layout_width="match_parent"
        android:layout_height="wrap_content"
        android:layout_below="@id/id_button_2"
        android:text="@string/text_button_3" />

</RelativeLayout>
```

然后修改 res/values/strings.xml 文件,在其中定义几个在布局文件中用到的字符串引用,修改后的文件内容如下:

```xml
<?xml version="1.0" encoding="utf-8"?>
<resources>

    <string name="app_name">Ex12Background03</string>
    <string name="hello_world">Hello world!</string>

    <string name="text_button_1">播放1号音乐</string>
    <string name="text_button_2">播放2号音乐</string>
    <string name="text_button_3">停止播放音乐</string>

</resources>
```

现在编写用于播放音乐的 Service 程序代码。为了便于管理,在 src 工程目录下,新建一个名为 com.ttt.ex12background03.service 的包,在这个包下新建一个 MyService 类,修改 MyService.java 文件内容为如下代码:

```java
package com.ttt.ex12background03.service;

import java.io.IOException;

import com.ttt.ex12background03.MainActivity;

import android.app.Notification;
import android.app.NotificationManager;
import android.app.PendingIntent;
import android.app.Service;
import android.content.Intent;
import android.media.AudioManager;
import android.media.MediaPlayer;
import android.media.MediaPlayer.OnCompletionListener;
import android.media.MediaPlayer.OnErrorListener;
import android.media.MediaPlayer.OnPreparedListener;
import android.os.Environment;
```

```
import android.os.IBinder;
import android.widget.Toast;

public class MyService extends Service {
    private MediaPlayer mp;

    private NotificationManager mNM;
    private int mNOTIFICATION = (int)System.currentTimeMillis();

    @Override
    public void onCreate() {
        mp = null;
        mNM = (NotificationManager)getSystemService(NOTIFICATION_SERVICE);
        showNotification();

        System.out.println("onCreate called");
    }

    @Override
    public int onStartCommand(Intent intent, int flags, int startId) {
        System.out.println("onStartCommand called");

        if (mp != null) {
            if (mp.isPlaying() == true) {
                mp.stop();
            }
            mp.release();
        }

        mp = new MediaPlayer();

        mp.setAudioStreamType(AudioManager.STREAM_MUSIC);
        mp.setOnPreparedListener(new OnPreparedListener() {
            @Override
            public void onPrepared(MediaPlayer mp) {
                mp.start();
            }
        });
        mp.setOnErrorListener(new OnErrorListener() {
            @Override
            public boolean onError(MediaPlayer mp, int what, int extra) {
                Toast.makeText(MyService.this, "无法播放该文件",
                                            Toast.LENGTH_LONG).show();
                return false;
            }
        });
        mp.setOnCompletionListener(new OnCompletionListener() {
            @Override
            public void onCompletion(MediaPlayer mp) {
                MyService.this.stopSelf();
            }
        });
```

```java
        try {
            String music = intent.getStringExtra("music");

            String sd_card_path =
                    Environment.getExternalStorageDirectory().getAbsolutePath();
            mp.setDataSource(sd_card_path + music);
            mp.prepareAsync();
        } catch (IllegalArgumentException e) {
            e.printStackTrace();
        } catch (SecurityException e) {
            e.printStackTrace();
        } catch (IllegalStateException e) {
            e.printStackTrace();
        } catch (IOException e) {
            e.printStackTrace();
        }

        return START_STICKY;
    }

    @Override
    public IBinder onBind(Intent intent) {
        // 由于我们不提供绑定式服务,因此,直接返回 null
        return null;
    }

    @Override
    public void onDestroy() {
        if (mp != null) {
            if (mp.isPlaying() == true) {
                mp.stop();
            }
            mp.release();
        }

        mNM.cancel(mNOTIFICATION);
        System.out.println("onDestroy called");
    }

    @SuppressWarnings("deprecation")
    private void showNotification() {
        CharSequence text = "Playing Music";
        Notification notification = new Notification(android.R.drawable.ic_media_play,
                text, System.currentTimeMillis());
        PendingIntent contentIntent = PendingIntent.getActivity(this, 0,
                new Intent(this, MainActivity.class), 0);
        notification.setLatestEventInfo(this, text, text + "……", contentIntent);
        mNM.notify(mNOTIFICATION, notification);
    }

}
```

在 MyService 的 onCreate()回调函数中,我们只是简单地初始化相关变量并在系统通知栏显示一个通知。在 onStartCommand()回调函数中，我们首先停止之前播放的音乐（如果之前正在播放的话），然后，设置播放的相关接口，特别需要注意 OnCompletionListener 接口，当这个接口函数被调用时，表示已经播放完指定的音乐，因此，在其处理函数中，我们停止被启动的服务，内容如下：

```
        mp.setOnCompletionListener(new OnCompletionListener() {
            @Override
            public void onCompletion(MediaPlayer mp) {
                MyService.this.stopSelf();
            }
        });
```

还需要特别说明 onStartCommand()回调函数的返回值的含义。onStartCommand()函数的返回值只能是如下的几个值之一，这些值及其含义如下。

（1）START_NOT_STICKY。

这个返回值告诉系统，当系统资源紧缺时，在 onStartCommand()函数返回后，如果系统结束了这个服务，则不需要重新创建和启动这个服务。

（2）START_STICKY。

这个返回值告诉系统，当系统资源紧缺时，在 onStartCommand()函数返回后，如果系统结束了这个服务，则需要重新创建和启动这个服务，并且再次使用 Intent 为 null 的参数调用服务的 onStartCommand()回调函数。

（3）START_REDELIVER_INTENT。

这个返回值告诉系统，当系统资源紧缺时，在 onStartCommand()函数返回后，如果系统结束了这个服务，则需要重新创建和启动这个服务，并且使用之前调用 onStartCommand()时的 Intent 参数再次调用服务的 onStartCommand()回调函数。

不管采用何种方式结束服务，系统都将调用 MyService 的 onDestroy()回调函数。在这里我们停止正在播放的音乐，并取消在通知栏中显示的通知。

现在修改 MainActivity 类，修改后的程序代码如下：

```
package com.ttt.ex12background03;

import com.ttt.ex12background03.service.MyService;

import android.support.v7.app.AppCompatActivity;
import android.content.Intent;
import android.os.Bundle;
import android.view.Menu;
import android.view.MenuItem;
import android.view.View;
import android.view.View.OnClickListener;
import android.widget.Button;

public class MainActivity extends AppCompatActivity implements OnClickListener {

    @Override
    protected void onCreate(Bundle savedInstanceState) {
        super.onCreate(savedInstanceState);
        setContentView(R.layout.activity_main);
```

```java
        Button btn1 = (Button)this.findViewById(R.id.id_button_1);
        btn1.setOnClickListener(this);
        Button btn2 = (Button)this.findViewById(R.id.id_button_2);
        btn2.setOnClickListener(this);
        Button btn3 = (Button)this.findViewById(R.id.id_button_3);
        btn3.setOnClickListener(this);
    }

    @Override
    public void onClick(View v) {
        int id = v.getId();

        if (id == R.id.id_button_1) {
            Intent service = new Intent(this, MyService.class);
            service.putExtra("music", "/MyMusic/AnniInMirror.mp3");
            this.startService(service);
        }
        else if (id == R.id.id_button_2) {
            Intent service = new Intent(this, MyService.class);
            service.putExtra("music", "/short1.mp3");
            this.startService(service);
        }
        else {
            System.out.println("Stopping Service");
            Intent service = new Intent(this, MyService.class);
            this.stopService(service);
        }
    }
}
```

在这个类中，我们显示主界面，并设置对按钮点击的响应函数。注意对按钮点击的响应处理 onClick()函数。在这里，当点击第一个或第二个按钮时，我们创建一个用于启动 Service 的 Intent 对象，并传递一个需要播放的音乐文件名，然后调用 startService()来启动指定的服务运行。当点击第三个按钮时，我们调用 stopService()来结束指定的服务。

在运行程序之前，我们需要在 SD 卡上放置两个音乐文件：/short1.mp3 和/MyMusic/AnniInMirror.mp3。并且，还需要修改 AndroidManifest.xml 文件，在其中声明服务程序 MyService。修改后的 AndroidManifest.xml 文件内容如下：

```xml
<?xml version="1.0" encoding="utf-8"?>
<manifest xmlns:android="http://schemas.android.com/apk/res/android"
    package="com.ttt.ex12background03">

    <application
        android:allowBackup="true"
        android:icon="@drawable/ic_launcher"
        android:label="@string/app_name"
        android:theme="@style/AppTheme" >
        <activity
            android:name=".MainActivity"
            android:label="@string/app_name" >
            <intent-filter>
                <action android:name="android.intent.action.MAIN" />
```

```xml
            <category android:name="android.intent.category.LAUNCHER" />
        </intent-filter>
    </activity>

    <service android:name="com.ttt.ex12background03.service.MyService"/>

</application>
</manifest>
```

其中的

```xml
    <service android:name="com.ttt.ex12background03.service.MyService"/>
```

就声明了需要使用的 Service。

运行程序，即可得到如图 12-5 所示的结果。

12.5 同步练习二

Android 提供了一个实用的 Service 子类，称为 IntentService，使用 IntentService 可以简化服务程序的编写内容。自学 IntentService，并使用 IntentService 完成时钟实时显示的练习程序。
提示：重写 InentService 的 onHandleIntent()方法，通过延时操作，修改界面上时钟的显示状态。

使用网络

在互联网时代,几乎没有一个 Android 应用程序是独立于网络之外的,因此,开发基于网络的应用无疑是必要的。本章对在 Android 程序中如何使用网络进行介绍,包括如下内容:使用 ConnectivityManager 管理网络状态、使用 HttpURLConnection 访问网络、使用 OkHttp 访问网络,同时,我们还对如何使用 JSON(JavaScript Object Notation)格式的数据与后台服务进行通信做一个简单的介绍。

13.1 使用 ConnectivityManager 管理网络状态

在使用网络进行任何数据通信之前,首先需要获得网络的状态,例如,当前网络是否开启,是 Wi-Fi 连接、GPRS 连接还是 UMTS 连接等。我们需要使用 Android SDK 提供的 ConnectivityManager 类来获得网络的状态。想要获得一个 ConnectivityManager 类的对象,需要使用如下语句:

```
ConnectivityManager cm = (ConnectivityManager)
        Context.getSystemService(Context.CONNECTIVITY_SERVICE);
```

一旦得到 ConnectivityManager 类的对象,我们可以使用它提供的各种方法和属性来检查网络状态和监听网络状态的变化。ConnectivityManager 常用的方法如下。

(1) public NetworkInfo getActiveNetworkInfo()。

获得当前活动的网络信息。如果当前没有活动的网络,则返回 null。

(2) public NetworkInfo[] getAllNetworkInfo()。

获得系统支持的所有网络类型的信息。

如果想要获得及监听网络状态的变化,应用程序需要在 AndroidManifest.xml 文件中申请 android.permission.ACCESS_NETWORK_STATE 权限。

下面我们举一个简单的例子来说明 ConnectivityManager 类的使用方法。这个例子首先检查当前是否有活动的网络,如果有,则打印出这个活动网络的相关信息。运行结果如图 13-1 所示。

图 13-1 检查当前是否有活动网络的程序运行结果

由于程序是在模拟器上运行的,而模拟器的 mobile 数据网络是开启的,因此,程序在 TextView 中显示 mobile 数据网络是可用的。

下面来构建这个程序。新建一个名为 Ex13Network01 的 Android 工程。修改 res/layout/activity_main.xml 文件,为其中的 TextView 组件添加一个 id 属性,修改后的文件内容如下:

```xml
<RelativeLayout xmlns:android="http://schemas.android.com/apk/res/android"
    android:layout_width="match_parent"
    android:layout_height="match_parent">

    <TextView
        android:id="@+id/id_textview"
        android:layout_width="wrap_content"
        android:layout_height="wrap_content"
        android:text="@string/hello_world" />

</RelativeLayout>
```

再修改 MainActivity.java 文件,修改后的代码如下:

```java
package com.ttt.ex13network01;

import android.support.v7.app.AppCompatActivity;
import android.net.ConnectivityManager;
import android.net.NetworkInfo;
import android.os.Bundle;
import android.widget.TextView;

public class MainActivity extends AppCompatActivity {

    @Override
    protected void onCreate(Bundle savedInstanceState) {
        super.onCreate(savedInstanceState);
        setContentView(R.layout.activity_main);

        StringBuffer sb = new StringBuffer();

        ConnectivityManager cm = (ConnectivityManager)
                    this.getSystemService(MainActivity.CONNECTIVITY_SERVICE);
        NetworkInfo ni = cm.getActiveNetworkInfo();
        if (ni == null) {
            sb.append("当前没有活动网络。");
        }
        else {
            if (ni.isConnected()){
                sb.append(ni.getTypeName()).append("是活动的。");
            }
            else {
                sb.append(ni.getTypeName()).append("不在服务区。");
            }
        }

        TextView tv = (TextView)this.findViewById(R.id.id_textview);
        tv.setText(sb.toString());
    }
```

}

在该类的 onCreate()回调函数中，显示界面，并获得 ConnectivityManager 对象，并调用 ConnectivityManager 的 getActiveNetworkInfo()获得当前活动网络的 NetworkInfo 对象，如果这个函数返回 null，则表示当前没有活动的数据网络，否则表示当前有活动的数据网络，但是为了保证网络是可用的，也就是当前网络是可以进行数据通信的，我们进一步调用该 NetworkInfo 的 isConnected()方法判断当前网络是否可用，并据此来显示相应的信息。

想要运行该程序，还需要在 AndroidManifest.xml 文件中添加如下的权限申请：

```
<uses-permission
        android:name="android.permission.ACCESS_NETWORK_STATE"/>
```

运行该程序，即可得到如图 13-1 所示的界面。

13.2 使用 HttpURLConnection 访问网络

使用 HttpURLConnection 是最直接与服务器进行基于 HTTP 通信的方式。使用 HttpURLConnection 与后台进行通信的编程过程如下。

（1）通过 URL.openConnection()得到一个 HttpURLConnection 对象。

（2）设置请求头相关的参数。

（3）在使用 POST 方法的请求中，调用 setDoOutput(true)，并通过 getOutputStream()得到输出流，进而向该输出流输出数据。

（4）处理响应数据。

（5）调用 disconnect()关闭连接。

HttpURLConnection 支持 HTTP 中规定的所有请求方式，基于上面介绍的编程过程，可以使用 GET、POST、HEAD、OPTION、DELETE、TRACE 方法向服务器发送请求。但是最常用的是 GET 方法和 POST 方法，下面我们对使用 GET 方法和 POST 方法发送请求并处理响应数据进行介绍。

为了仔细观察 Android 与服务器进行通信的过程，我们编写一个服务器端程序，使用 Servlet 来处理来自客户端的请求。该服务器端程序完成如下任务：接收来自 HTTP 的 GET 和 POST 请求，并根据请求参数 type 和 id 向请求客户端发送指定的图片数据。具体来说就是：客户端程序通过 GET 或 POST 向服务器端发送请求，其中的 type 参数指明请求的图片的类型，type=1 表示请求蝴蝶图片，type=2 表示请求卡通图片；id 参数指明请求的图片的编号，id=1 或 id=2 表示不同类型的两张图片。该 Servlet 的代码如下：

```
package com.ttt.servlet;

import java.io.FileInputStream;
import java.io.IOException;

import javax.servlet.ServletException;
import javax.servlet.ServletOutputStream;
import javax.servlet.annotation.WebServlet;
import javax.servlet.http.HttpServlet;
import javax.servlet.http.HttpServletRequest;
```

```
import javax.servlet.http.HttpServletResponse;

@WebServlet("/ImageShower")
public class ImageShower extends HttpServlet {
   private static final long serialVersionUID = 1L;

   protected void doGet(HttpServletRequest request, HttpServletResponse response)
                   throws ServletException, IOException {
      String type = request.getParameter("type");
      if ((type == null) || (type.equalsIgnoreCase(""))) {
         type = "1";
      }
      String id = request.getParameter("id");
      if ((id == null) || (id.equalsIgnoreCase(""))) {
         id = "1";
      }

      FileInputStream fis = new FileInputStream(
         this.getServletContext().getRealPath("") + "images/png" + type + id + ".png");
      byte[] b=new byte[fis.available()];
      fis.read(b);
      fis.close();

      response.setContentType("application/octet-stream");
      ServletOutputStream op = response.getOutputStream();
      op.write(b);
      op.close();
   }

   protected void doPost(HttpServletRequest request, HttpServletResponse response)
                   throws ServletException, IOException {
      doGet(request, response);
   }
}
```

4 张图片放置在 Web 应用的 images/pngxx.png 文件中，其中 xx 可以为 "11" "12" "21" "22"，4 张图片如图 13-2 所示。

图 13-2　4 张图片的效果

现在我们在 Android 客户端使用 HttpURLConnection 来获得并显示图片。我们分别对采用 GET 方法和 POST 方法来获取图片进行介绍。

13.2.1　使用 HttpURLConnection 的 GET 方法获取图片

我们首先使用 HttpURLConnection 的 GET 方法来获取服务器端的图片，并显示在手机上，程序运行效果如图 13-3 所示。

图 13-3　使用 GET 方法获取服务器端图片的程序首界面

点击 4 个按钮中的任意一个，将在手机的下方显示相应的图片，例如，如果点击第 3 个按钮，则显示如图 13-4 所示的界面。

图 13-4　点击按钮显示相应的图片

现在来构建这个程序。新建一个名为 Ex13Network02 的 Android 工程。修改 res/layout/activity_main.xml 布局文件，修改后的文件内容如下：

```xml
<LinearLayout xmlns:android="http://schemas.android.com/apk/res/android"
    android:layout_width="match_parent"
    android:layout_height="match_parent"
    android:orientation="vertical">

    <Button
        android:id="@+id/id_btn_1"
        android:layout_width="match_parent"
        android:layout_height="wrap_content"
        android:text="@string/text_btn_1" />

    <Button
        android:id="@+id/id_btn_2"
        android:layout_width="match_parent"
        android:layout_height="wrap_content"
        android:text="@string/text_btn_2" />
```

```xml
    <Button
        android:id="@+id/id_btn_3"
        android:layout_width="match_parent"
        android:layout_height="wrap_content"
        android:text="@string/text_btn_3" />

    <Button
        android:id="@+id/id_btn_4"
        android:layout_width="match_parent"
        android:layout_height="wrap_content"
        android:text="@string/text_btn_4" />

    <ImageView
        android:id="@+id/id_iv"
        android:layout_width="match_parent"
        android:layout_height="match_parent"
        android:scaleType="fitCenter"
        android:contentDescription="@string/hello_world"
        />

</LinearLayout>
```

这个布局文件很简单，只是显示几个 Button 和一个 ImageView 而已。然后，还需要修改 res/values/strings.xml 文件，在其中定义几个字符串引用，内容如下：

```xml
<?xml version="1.0" encoding="utf-8"?>
<resources>

    <string name="app_name">Ex13Network02</string>
    <string name="hello_world">Hello world!</string>

    <string name="text_btn_1">显示第一张蝴蝶</string>
    <string name="text_btn_2">显示第二张蝴蝶</string>
    <string name="text_btn_3">显示第一张卡通</string>
    <string name="text_btn_4">显示第二张卡通</string>

</resources>
```

现在修改 MainActivity.java 文件，修改后的文件内容如下：

```java
package com.ttt.ex13network02;

import java.io.BufferedInputStream;
import java.io.ByteArrayOutputStream;
import java.io.IOException;
import java.net.HttpURLConnection;
import java.net.MalformedURLException;
import java.net.URL;

import android.support.v7.app.AppCompatActivity;
import android.graphics.Bitmap;
import android.graphics.BitmapFactory;
import android.net.ConnectivityManager;
import android.net.NetworkInfo;
```

```java
import android.os.AsyncTask;
import android.os.Bundle;
import android.view.View;
import android.view.View.OnClickListener;
import android.widget.Button;
import android.widget.ImageView;
import android.widget.Toast;

public class MainActivity extends AppCompatActivity implements OnClickListener {
    private ImageView iv;

    @Override
    protected void onCreate(Bundle savedInstanceState) {
        super.onCreate(savedInstanceState);
        setContentView(R.layout.activity_main);

        iv = (ImageView) this.findViewById(R.id.id_iv);

        Button btn_1 = (Button) this.findViewById(R.id.id_btn_1);
        btn_1.setOnClickListener(this);
        Button btn_2 = (Button) this.findViewById(R.id.id_btn_2);
        btn_2.setOnClickListener(this);
        Button btn_3 = (Button) this.findViewById(R.id.id_btn_3);
        btn_3.setOnClickListener(this);
        Button btn_4 = (Button) this.findViewById(R.id.id_btn_4);
        btn_4.setOnClickListener(this);
    }

    @Override
    public void onClick(View v) {
        if (checkNetworkState() != true) {
            Toast.makeText(this, "网络没有打开,请打开网络后再试。",
                                    Toast.LENGTH_LONG).show();
            return;
        }

        int id = v.getId();
        switch (id) {
        case R.id.id_btn_1:
            downLoadImageAndShow(1, 1);
            break;
        case R.id.id_btn_2:
            downLoadImageAndShow(1, 2);
            break;
        case R.id.id_btn_3:
            downLoadImageAndShow(2, 1);
            break;
        case R.id.id_btn_4:
            downLoadImageAndShow(2, 2);
            break;
```

```java
        }
    }

    private boolean checkNetworkState() {
        ConnectivityManager cm = (ConnectivityManager) this
                .getSystemService(MainActivity.CONNECTIVITY_SERVICE);
        NetworkInfo ni = cm.getActiveNetworkInfo();
        if ((ni == null) || (ni.isConnected() == false)) {
            return false;
        }

        return true;
    }

    private void downLoadImageAndShow(int type, int id) {
        new MyAsyncTask().
            execute("http://172.18.171.253:8080/AImageShower/ImageShower?type=" +
            type + "&" + "id=" + id);
    }

    private class MyAsyncTask extends AsyncTask<String, Void, Bitmap> {
        @Override
        protected void onPreExecute() {
        }

        @Override
        protected void onProgressUpdate (Void... values) {
        }

        @Override
        protected void onPostExecute (Bitmap bm) {
            iv.setImageBitmap(bm);
        }

        @Override
        protected Bitmap doInBackground(String... params) {
            URL url;
            HttpURLConnection urlConnection = null;
            Bitmap bm = null;

            try {
                url = new URL(params[0]);
                urlConnection = (HttpURLConnection)url.openConnection();
                urlConnection.setRequestMethod("GET");

                urlConnection.connect();
                if (urlConnection.getResponseCode() != HttpURLConnection.HTTP_OK) {
                    urlConnection.disconnect();
                    return null;
                }

                byte[] b = new byte[2048];
```

```
                ByteArrayOutputStream baos = new ByteArrayOutputStream();

                BufferedInputStream in = new
                        BufferedInputStream(urlConnection.getInputStream());
            int len = 0;
                while((len = in.read(b))>0) {
                    baos.write(b, 0, len);
                }

                bm = BitmapFactory.decodeByteArray(baos.toByteArray(), 0, baos.size());
                baos.close();
                in.close();

            }catch (MalformedURLException e) {
                return null;
            }catch (IOException e) {
                return null;
            }
            finally {
                urlConnection.disconnect();
            }

            return bm;
        }
    }
}
```

在这个类的 onCreate()回调函数中，我们显示主界面，同时获得相关组件的引用，并设置对按钮点击的响应函数。在按钮点击的响应函数 onClick()中，首先判断当前是否有可用的网络，若有，则调用 downLoadImageAndShow()函数来下载图片并显示在 ImageView 组件中。

由于网络操作需要在独立的线程中进行，因此，在 downLoadImageAndShow()函数中，我们创建一个 AsyncTask 的子类 MyAsyncTask 类的对象来完成网络操作。在 MyAsyncTask 类的 doInBackground()函数中，我们首先定义一个指向服务器端 Servlet 的 URL 对象，打开这个网络连接并得到 HttpURLConnection 对象，然后设置请求方法为 GET 方法并发送请求，接下来检查服务器端程序是否正确处理了客户端的请求，若是，则接收从服务器端的 Servlet 发来的图片二进制流，通过 BitmapFactory 编码为 Bitmap 对象，然后将这个图片返回给 PostExcute()函数进而将图片显示在 ImageView 中。

想要运行该程序，我们还需要修改 AndroidManifest.xml 文件，对网络进行相应的授权申请，为此，修改 AndroidManifest.xml 文件，在其中添加如下代码：

```
<uses-permission android:name="android.permission.ACCESS_NETWORK_STATE"/>
<uses-permission android:name="android.permission.INTERNET"/>
```

运行该程序，即可得到如图 13-3 所示的界面。

13.2.2　使用 HttpURLConnection 的 POST 方法获取图片

现在我们演示如何使用 HttpURLConnection 的 POST 方法从服务器端获取图片，程序的运行效果与图 13-3 完全一样。我们只需要修改 downLoadImageAndShow()函数，使之使用 POST 方法来获得图片数据。修改后的 downLoadImageAndShow()函数如下：

```java
private void downLoadImageAndShow(int type, int id) {
    new MyAsyncTask(type,id).
        execute("http://172.18.171.253:8080/AImageShower/ImageShower");
}

private class MyAsyncTask extends AsyncTask<String, Void, Bitmap> {
    private int type, id;

    public MyAsyncTask(int type, int id) {
        this.type = type;
        this.id = id;
    }

    @Override
    protected void onPreExecute() {
    }

    @Override
    protected void onProgressUpdate (Void... values) {
    }

    @Override
    protected void onPostExecute (Bitmap bm) {
        iv.setImageBitmap(bm);
    }

    @Override
    protected Bitmap doInBackground(String... params) {
        URL url;
        HttpURLConnection urlConnection = null;
        Bitmap bm = null;

        try {
            url = new URL(params[0]);
            urlConnection = (HttpURLConnection)url.openConnection();

            urlConnection.setRequestMethod("POST");
            urlConnection.setDoInput(true);
            urlConnection.setDoOutput(true);

            urlConnection.setUseCaches(false);
            urlConnection.setInstanceFollowRedirects(true);
            urlConnection.setChunkedStreamingMode(0);
            urlConnection.setRequestProperty("Content-Type",
                            "application/x-www-form-urlencoded");

            urlConnection.connect();
            DataOutputStream out = new
                        DataOutputStream(urlConnection.getOutputStream());

            String content = "type=" + URLEncoder.encode("" + type, "UTF-8") +
                        "&id=" + URLEncoder.encode("" + id, "UTF-8");
```

```
            out.writeBytes(content);
            out.flush();
            out.close();

            if (urlConnection.getResponseCode() != HttpURLConnection.HTTP_OK) {
                urlConnection.disconnect();
                return null;
            }

            byte[] b = new byte[2048];
            ByteArrayOutputStream baos = new ByteArrayOutputStream();
            BufferedInputStream in = new
                        BufferedInputStream(urlConnection.getInputStream());
        int len = 0;
            while((len = in.read(b))>0) {
                baos.write(b, 0, len);
            }

            bm = BitmapFactory.decodeByteArray(baos.toByteArray(), 0, baos.size());
            baos.close();
            in.close();
    }catch (MalformedURLException e) {
        return null;
    }catch (IOException e) {
        return null;
    }
    finally {
        urlConnection.disconnect();
    }

    return bm;
    }
}
```

与采用 GET 方法获取图片不同的关键点在于如下代码：

```
            urlConnection.setRequestMethod("POST");
```

此代码用于设置访问方法为 POST；同时

```
            urlConnection.setDoInput(true);
            urlConnection.setDoOutput(true);
```

这两句是必需的，用于告诉 HttpURLConnection 对象我们将采用 HTTP 请求体来发送请求数据；同时如下代码：

```
            urlConnection.setChunkedStreamingMode(0);
```

表示为了加快数据传输的速度，设置每个数据块的大小为默认数据块大小；同时如下代码：

```
            urlConnection.setRequestProperty("Content-Type",
                        "application/x-www-form-urlencoded");
```

表示我们将采用表单的形式向服务器端程序发送请求数据；然后使用如下代码：

```
            urlConnection.connect();
            DataOutputStream out = new
                        DataOutputStream(urlConnection.getOutputStream());
```

```
            String content = "type=" + URLEncoder.encode("" + type, "UTF-8") +
                             "&id=" + URLEncoder.encode("" + id, "UTF-8");
            out.writeBytes(content);
            out.flush();
            out.close();
```

获得一个连接到服务器端的数据流,并向服务器端程序发送请求数据,注意,对于数据体部分我们采用了 UTF-8 的编码表示方案。其他的代码与 GET 方法类似,不再赘述。

运行修改后的程序,将得到与图 13-3 完全一样的效果。

现在我们还需要回答一个问题:既然既可以使用 GET 方法也可以使用 POST 方法来处理网络通信,那么在什么情况下使用 GET 及在什么情况下使用 POST 呢?回答是:当使用 GET 方法进行网络请求时,请求参数附加在 URL 地址的后面,这种方法简单但是不安全,且参数长度受到 2048 字节的限制;而使用 POST 方法进行网络请求,请求参数是放置在请求体中的,安全且不受请求参数长度的限制。

13.3 同步练习一

仿照 13.2 节中的例子,使用基于 Java 的多线程机制,也就是使用 Thread 机制完成与 13.2 节中的例子程序类似的功能。要求:在程序界面中选择使用 GET 方法或 POST 方法获取图片。

13.4 使用 OkHttp 访问网络

由于 Apache 的 HttpClient 非常庞大且复杂,因此,从 Android 5.1 版本开始,Google 的 Android 开发团队不再将 Apache 的 HttpClient 纳入 Android SDK 中,想要使用 HTTP,读者可以使用 13.2 节介绍的 HttpUrlConnection 类或使用第三方开源的 http 包。在此,我们使用完善且易用的 OkHttp。

13.4.1 使用 Get 方法进行服务请求

想要使用 OkHttp,我们需要从 OkHttp 官网上下载最新的 OkHttp 包和 Okio 包,将它们放置到工程的 app/libs 目录下,对每个包,选择该包并单击鼠标右键,在弹出的快捷菜单中选择"Add as Library"。做好了这些准备工作,就可以使用 OkHttp 了,基本过程如下。

(1) 创建 OkHttpClient 对象,代码如下:

```
private final OkHttpClient client = new OkHttpClient();
```

(2) 创建 http 请求,代码如下:

```
Request request = new Request.Builder()
        .url("http://publicobject.com/helloworld.txt")
        .build();
```

(3) 发送请求到服务器端,代码如下:

```
Response response = client.newCall(request).execute();
```

(4) 处理从服务器端返回的结果,代码如下:

```
if (!response.isSuccessful()) throw new IOException("Unexpected code " + response);
Headers responseHeaders = response.headers();
 for (int i = 0; i < responseHeaders.size(); i++) {
```

```
        System.out.println(responseHeaders.name(i) + ": " + responseHeaders.value(i));
    }
System.out.println(response.body().string());
```

在上面的第（3）步和第（4）步中，我们使用了同步方式来发送请求和接收响应，也可以使用异步方式来发送请求和接收响应。因此，上面的第（3）步和第（4）步可替换为如下代码：

```
client.newCall(request).enqueue(new Callback() {
    @Override public void onFailure(Request request, Throwable throwable) {
      throwable.printStackTrace();
    }

    @Override public void onResponse(Response response) throws IOException {
        if (!response.isSuccessful()) throw new IOException("Unexpected code " +
                                                            response);

        Headers responseHeaders = response.headers();
    for (int i = 0; i < responseHeaders.size(); i++) {
            System.out.println(responseHeaders.name(i) + ": " +
                                            responseHeaders.value(i));
        }
        System.out.println(response.body().string());
    }
});
```

13.4.2 使用 Post 方法进行服务请求

使用 Post 方法进行服务请求可以包括普通字符串、文件、表单等多种格式的请求数据到服务器端程序，并且这些请求数据都是打包到请求包的请求体中的，所以相比于 Get 请求，相对安全和灵活一些。

1. 使用 Post 方法提交 String

使用 Post 方法提交 String 的实现代码如下所示：

```
public static final MediaType MEDIA_TYPE_MARKDOWN
                        = MediaType.parse("text/x-markdown; charset=utf-8");
private final OkHttpClient client = new OkHttpClient();

public void postString() throws Exception {
    String postBody = ""
            + "Releases\n"
            + "--------\n"
            + "\n"
            + " * _1.0_ May 6, 2013\n"
            + " * _1.1_ June 15, 2013\n"
            + " * _1.2_ August 11, 2013\n";

    Request request = new Request.Builder()
                .url("https://api.github.com/markdown/raw")
                .post(RequestBody.create(MEDIA_TYPE_MARKDOWN, postBody))
                .build();
    Response response = client.newCall(request).execute();
    if (!response.isSuccessful()) throw new IOException("Unexpected code " + response);
     System.out.println(response.body().string());
}
```

2. 使用 Post 方法提交文件

使用 Post 方法提交文件的实现代码如下所示：

```java
public static final MediaType MEDIA_TYPE_MARKDOWN
                = MediaType.parse("text/x-markdown; charset=utf-8");
private final OkHttpClient client = new OkHttpClient();

public void postFile() throws Exception {
    File file = new File("README.md");
    Request request = new Request.Builder()
            .url("https://api.github.com/markdown/raw")
            .post(RequestBody.create(MEDIA_TYPE_MARKDOWN, file))
            .build();
    Response response = client.newCall(request).execute();
    if (!response.isSuccessful()) throw new IOException("Unexpected code " + response);
    System.out.println(response.body().string());
}
```

3. 使用 Post 方法提交简单表单

使用 FormEncodingBuilder 来构建与 HTML<form>标签相同效果的请求体，"key/value"对将使用一种 HTML 兼容形式的 URL 编码来进行编码，代码如下：

```java
private final OkHttpClient client = new OkHttpClient();

public void postForm() throws Exception {
    RequestBody formBody = new FormBody.Builder()
                    .add("search", "Jurassic Park")
                    .build();
    Request request = new Request.Builder()
                    .url("https://en.wikipedia.org/w/index.php")
                    .post(formBody)
                    .build();
    Response response = client.newCall(request).execute();
    if (!response.isSuccessful()) throw new IOException("Unexpected code " + response);
    System.out.println(response.body().string());
}
```

4. 使用 Post 方法提交分块（Multipart）表单

MultipartBody.Builder 可以构建复杂的请求体，与 HTML 文件上传形式兼容，多块请求体中每块请求都是一个请求体，可以定义自己的请求头。这些请求头可以用来描述这些块请求，如 Content-Disposition。如果 Content-Length 和 Content-Type 可用的话，它们会被自动添加到请求头中，代码如下：

```java
private static final String IMGUR_CLIENT_ID = "……";
private static final MediaType MEDIA_TYPE_PNG = MediaType.parse("image/png");
private final OkHttpClient client = new OkHttpClient();

public void postMultiPartForm() throws Exception {
    RequestBody requestBody = new MultipartBody.Builder()
                    .setType(MultipartBody.FORM)
                    .addFormDataPart("title", "Square Logo")
```

```
                    .addFormDataPart("image", "logo-square.png",
                        RequestBody.create(MEDIA_TYPE_PNG, new
                        File("website/static/logo-square.png")))
                    .build();
    Request request = new Request.Builder()
                    .header("Authorization", "Client-ID " + IMGUR_CLIENT_ID)
                    .url("https://api.baidu.com/3/image")
                    .post(requestBody)
                    .build();
    Response response = client.newCall(request).execute();
    if (!response.isSuccessful()) throw new IOException("Unexpected code " + response);
    System.out.println(response.body().string());
}
```

13.4.3　设置请求头及提取响应头

典型的 HTTP 头是一个 Map<String, String>：每个字段都有一个值或没有值，也有一些头允许有多个值。

当写请求头的时候，使用 header(name, value)可以设置唯一的"name/value"对。如果已经有值，旧的值将被移除，然后再添加新的值。使用 addHeader(name, value)可以添加多个值。

当读取响应头时，使用 header(name)返回最后出现的"name/value"对。通常情况这也是唯一的"name/value"对。如果没有值，那么 header(name)将返回 null，如果想读取字段对应的所有值，则使用 headers(name)会返回一个列表。为了获取所有的 Header，Headers 类支持按 index 方式访问，代码如下：

```
private final OkHttpClient client = new OkHttpClient();
public void headerExample() throws Exception {
    Request request = new Request.Builder()
            .url("https://api.github.com/repos/square/okhttp/issues")
            .header("User-Agent", "OkHttp Headers.java")
            .addHeader("Accept", "application/json; q=0.5")
            .addHeader("Accept", "application/vnd.github.v3+json")
            .build();
    Response response = client.newCall(request).execute();
    if (!response.isSuccessful()) throw new IOException("Unexpected code " + response);
    System.out.println("Server: " + response.header("Server"));
    System.out.println("Date: " + response.header("Date"));
    System.out.println("Vary: " + response.headers("Vary"));
}
```

13.4.4　配置 OkHttp 超时

没有响应时使用超时结束 call。没有响应的原因可能是客户端链接问题、服务器可用性问题，或者这之间的其他问题等。OkHttp 支持连接超时、读取超时和写入超时。声明 client 类型的代码如下：

```
private final OkHttpClient client;

public ConfigureTimeouts() throws Exception {
    client = new OkHttpClient.Builder()
            .connectTimeout(10, TimeUnit.SECONDS)
            .writeTimeout(10, TimeUnit.SECONDS)
```

```
            .readTimeout(30, TimeUnit.SECONDS)
            .build();
}

public void run() throws Exception {
    Request request = new Request.Builder().url("http://httpbin.org/delay/2").build();
    Response response = client.newCall(request).execute();
    System.out.println("Response completed: " + response);
}
```

13.5　OkHttp Get 实现示例

下面我们通过一个具体的例子来演示如何使用 OkHttp 的 Get 方法（也就是 HTTP 的 GET 方法请求）来与服务器端程序进行 HTTP 通信。继续使用 13.2 节中的点击按钮显示相应图片的例子。程序运行效果与图 13-3 完全一样。

现在我们来构建这个程序。新建一个名为 Ex13Network03 的 Android 工程。修改 res/layout/activity_main.xml 文件，修改后的文件内容如下：

```xml
<?xml version="1.0" encoding="utf-8"?>
<LinearLayout xmlns:android="http://schemas.android.com/apk/res/android"
    android:id="@+id/activity_main"
    android:layout_width="match_parent"
    android:layout_height="match_parent"
    android:orientation="vertical">

    <Button
        android:id="@+id/id_btn_1"
        android:layout_width="match_parent"
        android:layout_height="wrap_content"
        android:text="@string/text_btn_1" />

    <Button
        android:id="@+id/id_btn_2"
        android:layout_width="match_parent"
        android:layout_height="wrap_content"
        android:text="@string/text_btn_2" />

    <Button
        android:id="@+id/id_btn_3"
        android:layout_width="match_parent"
        android:layout_height="wrap_content"
        android:text="@string/text_btn_3" />

    <Button
        android:id="@+id/id_btn_4"
        android:layout_width="match_parent"
        android:layout_height="wrap_content"
        android:text="@string/text_btn_4" />

    <ImageView
        android:id="@+id/id_iv"
```

```xml
        android:layout_width="match_parent"
        android:layout_height="match_parent"
        android:scaleType="fitCenter"
        android:contentDescription="@string/hello_world"
        />

</LinearLayout>
```

这个布局文件很简单，只是显示了几个 Button 和一个 ImageView 而已。现在修改 res/values/strings.xml 文件，在其中定义几个在布局文件中用到的字符串引用，内容如下：

```xml
<?xml version="1.0" encoding="utf-8"?>
<resources>

    <string name="app_name">Ex13Network03</string>
    <string name="hello_world">Hello world!</string>

    <string name="text_btn_1">显示第一张蝴蝶</string>
    <string name="text_btn_2">显示第二张蝴蝶</string>
    <string name="text_btn_3">显示第一张卡通</string>
    <string name="text_btn_4">显示第二张卡通</string>

</resources>
```

修改 MainActivity.java 文件，使之显示主界面，处理对按钮的点击事件，并根据点击的按钮通过 OkHttp 的 Get 方法从服务器端获得相应的图片并显示在 ImageView 组件中，修改后的文件内容如下：

```java
package com.ttt.ex13network03;

import android.graphics.Bitmap;
import android.graphics.BitmapFactory;
import android.net.ConnectivityManager;
import android.net.NetworkInfo;
import android.os.Bundle;
import android.os.Handler;
import android.support.design.widget.FloatingActionButton;
import android.support.design.widget.Snackbar;
import android.support.v7.app.AppCompatActivity;
import android.support.v7.widget.Toolbar;
import android.view.View;
import android.widget.Button;
import android.widget.ImageView;
import android.widget.Toast;

import java.io.IOException;

import okhttp3.OkHttpClient;
import okhttp3.Request;
import okhttp3.Response;

public class MainActivity extends AppCompatActivity implements View.OnClickListener {
    private ImageView iv;
    private Bitmap bm;
```

```java
private Handler handler;

@Override
protected void onCreate(Bundle savedInstanceState) {
    super.onCreate(savedInstanceState);
    setContentView(R.layout.activity_main);

    bm = null;
    handler = new Handler();

    iv = (ImageView) this.findViewById(R.id.id_iv);

    Button btn_1 = (Button) this.findViewById(R.id.id_btn_1);
    btn_1.setOnClickListener(this);
    Button btn_2 = (Button) this.findViewById(R.id.id_btn_2);
    btn_2.setOnClickListener(this);
    Button btn_3 = (Button) this.findViewById(R.id.id_btn_3);
    btn_3.setOnClickListener(this);
    Button btn_4 = (Button) this.findViewById(R.id.id_btn_4);
    btn_4.setOnClickListener(this);
}

private boolean checkNetworkState() {
    ConnectivityManager cm = (ConnectivityManager) this
            .getSystemService(MainActivity.CONNECTIVITY_SERVICE);
    NetworkInfo ni = cm.getActiveNetworkInfo();
    if ((ni == null) || (!ni.isConnected())) {
        return false;
    }

    return true;
}

@Override
public void onClick(View v) {
    if (!checkNetworkState()) {
        Toast.makeText(this, "网络没有打开,请打开网络后再试。",
                Toast.LENGTH_LONG).show();
        return;
    }

    int id = v.getId();

    switch(id) {
        case R.id.id_btn_1:
            downloadImageAndShow(1, 1);
            break;
        case R.id.id_btn_2:
            downloadImageAndShow(1, 2);
            break;
        case R.id.id_btn_3:
            downloadImageAndShow(2, 1);
            break;
```

```
            case R.id.id_btn_4:
                downloadImageAndShow(2, 2);
                break;
        }
    }

    private void downloadImageAndShow(final int type, final int id) {
        new Thread(new Runnable() {
            @Override
            public void run() {
                OkHttpClient client = new OkHttpClient();

                Request request = new Request.Builder()
                        .url("http://192.168.233.131:8080/ForAndroid/ImageShower?" +
                            "type=" + type + "&id=" + id)
                        .build();

                Response response;
                try {
                    response = client.newCall(request).execute();
                } catch (IOException e) {
                    e.printStackTrace();
                    return;
                }
                if (!response.isSuccessful()) {
                    return;
                }

                byte[] b = null;
                try {
                    b = response.body().bytes();
                } catch (IOException e) {
                    e.printStackTrace();
                    return;
                }
                bm = BitmapFactory.decodeByteArray(b, 0, b.length);
                handler.post(new Runnable(){
                    @Override
                    public void run() {
                        iv.setImageBitmap(bm);
                    }
                });
            }
        }).start();
    }
}
```

在对按钮点击处理 onClick()函数中，程序根据所点击的按钮，调用 downloadImageAndShow()函数来获得相应的图片并显示。在 downloadImageAndShow()函数中，我们使用 Java 的 Thread 线程机制来创建获得图片的线程。在运行程序之前，需要在 AndroidManifest.xml 文件中对网络相关的调用进行授权，代码如下：

```
<uses-permission android:name="android.permission.ACCESS_NETWORK_STATE"/>
```

```
<uses-permission android:name="android.permission.INTERNET"/>
```

运行该程序，即可得到与图 13-3 完全一样的效果。

13.6 OkHttp Post 实现示例

与 Get 方法将请求参数放置在 URL 地址中不同的是，HttpPost 将发送给服务器端程序的请求参数放置在请求数据体中。理论上，采用 Post 方法，可以发送任何合法格式的数据给服务器端程序。

下面我们通过一个具体的例子来演示如何使用 OkHttp 的 Post 方法（也就是 HTTP 的 POST 方法请求）来与服务器端程序进行 HTTP 通信。继续使用 13.2 节中的点击按钮显示相应图片的例子。程序运行效果与图 13-3 完全一样。

现在我们来构建这个程序。新建一个名为 Ex13Network04 的 Android 工程。修改 res/layout/activity_main.xml 文件，修改后的文件内容如下：

```xml
<LinearLayout xmlns:android="http://schemas.android.com/apk/res/android"
    android:layout_width="match_parent"
    android:layout_height="match_parent"
    android:orientation="vertical">

    <Button
        android:id="@+id/id_btn_1"
        android:layout_width="match_parent"
        android:layout_height="wrap_content"
        android:text="@string/text_btn_1" />

    <Button
        android:id="@+id/id_btn_2"
        android:layout_width="match_parent"
        android:layout_height="wrap_content"
        android:text="@string/text_btn_2" />

    <Button
        android:id="@+id/id_btn_3"
        android:layout_width="match_parent"
        android:layout_height="wrap_content"
        android:text="@string/text_btn_3" />

    <Button
        android:id="@+id/id_btn_4"
        android:layout_width="match_parent"
        android:layout_height="wrap_content"
        android:text="@string/text_btn_4" />

    <ImageView
        android:id="@+id/id_iv"
        android:layout_width="match_parent"
        android:layout_height="match_parent"
        android:scaleType="fitCenter"
        android:contentDescription="@string/hello_world"
```

```
    />

</LinearLayout>
```

这个布局文件很简单,只是显示了几个 Button 和一个 ImageView 而已。现在修改 res/values/strings.xml 文件,在其中定义几个在布局文件中用到的字符串引用,内容如下:

```xml
<?xml version="1.0" encoding="utf-8"?>
<resources>

    <string name="app_name">Ex13Network04</string>
    <string name="hello_world">Hello world!</string>

    <string name="text_btn_1">显示第一张蝴蝶</string>
    <string name="text_btn_2">显示第二张蝴蝶</string>
    <string name="text_btn_3">显示第一张卡通</string>
    <string name="text_btn_4">显示第二张卡通</string>

</resources>
```

修改 MainActivity.java 文件,使之显示主界面,处理对按钮的点击事件,并根据点击的按钮通过 OKHttp 的 Post 方法从服务器端程序获得相应的图片并显示在 ImageView 组件中,修改后的文件内容如下:

```java
package com.ttt.ex13network04;

import android.graphics.Bitmap;
import android.graphics.BitmapFactory;
import android.net.ConnectivityManager;
import android.net.NetworkInfo;
import android.os.Bundle;
import android.os.Handler;
import android.support.v7.app.AppCompatActivity;
import android.view.View;
import android.widget.Button;
import android.widget.ImageView;
import android.widget.Toast;

import java.io.IOException;

import okhttp3.Call;
import okhttp3.Callback;
import okhttp3.FormBody;
import okhttp3.Headers;
import okhttp3.MediaType;
import okhttp3.OkHttpClient;
import okhttp3.Request;
import okhttp3.RequestBody;
import okhttp3.Response;

public class MainActivity extends AppCompatActivity implements View.OnClickListener {
    private ImageView iv;
    private Bitmap bm;
    private Handler handler;

    @Override
```

```java
protected void onCreate(Bundle savedInstanceState) {
    super.onCreate(savedInstanceState);
    setContentView(R.layout.activity_main);

    bm = null;
    handler = new Handler();

    iv = (ImageView) this.findViewById(R.id.id_iv);

    Button btn_1 = (Button) this.findViewById(R.id.id_btn_1);
    btn_1.setOnClickListener(this);
    Button btn_2 = (Button) this.findViewById(R.id.id_btn_2);
    btn_2.setOnClickListener(this);
    Button btn_3 = (Button) this.findViewById(R.id.id_btn_3);
    btn_3.setOnClickListener(this);
    Button btn_4 = (Button) this.findViewById(R.id.id_btn_4);
    btn_4.setOnClickListener(this);
}

private boolean checkNetworkState() {
    ConnectivityManager cm = (ConnectivityManager) this
            .getSystemService(MainActivity.CONNECTIVITY_SERVICE);
    NetworkInfo ni = cm.getActiveNetworkInfo();
    if ((ni == null) || (!ni.isConnected())) {
        return false;
    }

    return true;
}

@Override
public void onClick(View v) {
    if (!checkNetworkState()) {
        Toast.makeText(this, "网络没有打开,请打开网络后再试。",
                Toast.LENGTH_LONG).show();
        return;
    }

    int id = v.getId();

    switch(id) {
        case R.id.id_btn_1:
            downloadImageAndShow(1, 1);
            break;
        case R.id.id_btn_2:
            downloadImageAndShow(1, 2);
            break;
        case R.id.id_btn_3:
            downloadImageAndShow(2, 1);
            break;
        case R.id.id_btn_4:
            downloadImageAndShow(2, 2);
            break;
    }
}
```

```java
    private void downloadImageAndShow(final int type, final int id) {
        new Thread(new Runnable() {
            @Override
            public void run() {
                OkHttpClient client = new OkHttpClient();
                RequestBody formBody = new FormBody.Builder()
                        .add("type", "" + type)
                        .add("id", "" + id)
                        .build();
                Request request = new Request.Builder()
                        .url("http://192.168.233.131:8080/ForAndroid/ImageShower")
                        .post(formBody)
                        .build();

                Response response;
                client.newCall(request).enqueue(new Callback() {
                    @Override
                    public void onFailure(Call call, IOException e) {
                        e.printStackTrace();
                    }

                    @Override
                    public void onResponse(Call call, Response response)
                                                    throws IOException {
                        if (!response.isSuccessful()) {
                            return;
                        }

                        byte[] b = null;
                        try {
                            b = response.body().bytes();
                        } catch (IOException e) {
                            e.printStackTrace();
                            return;
                        }
                        bm = BitmapFactory.decodeByteArray(b, 0, b.length);
                        handler.post(new Runnable() {
                            @Override
                            public void run() {
                                iv.setImageBitmap(bm);
                            }
                        });
                    }
                });
            }
        }).start();
    }
}
```

在对按钮点击处理 onClick()函数中，程序根据所点击的按钮，调用 downloadImageAndShow()函数来获得相应的图片并显示。在 downloadImageAndShow()函数中，我们使用 Java 的 Thread 线程机制来创建获得图片的线程。在这段代码中，我们还使用了 OkHttp 的异步机制来获取图片。

在运行程序之前，需要在 AndroidManifest.xml 文件中对网络相关的调用进行授权，代码如下：

```
<uses-permission android:name="android.permission.ACCESS_NETWORK_STATE"/>
<uses-permission android:name="android.permission.INTERNET"/>
```

运行该程序，即可得到与图 13-3 完全一样的效果。

13.7 同步练习二

使用 HttpClient 从任意一个公共网站上使用 Get 方法请求一个 HTML 页面，并将得到的 HTML 页面显示在 WebView 组件中。

13.8 使用 Multipart 传递请求数据到服务器端程序

使用 Multipart Form 可以传递包含文件流的请求数据到服务器端，例如，在一个注册程序中，如果需要传递包括头像在内的注册信息到服务器端程序，则需要使用 Multipart 请求体。

下面我们通过一个例子来介绍如何使用 Multipart 格式从 Android 程序向服务器端程序传递 Multipart 数据体格式的请求数据，例子的运行效果与图 13-3 完全一样，只是实现的方式不同，同时，为了演示 Multipart 的数据请求，我们在请求数据包中附加了一个文件：一个简单的图片文件，服务器端程序收到这个图片后保存在其工作目录的 images 目录下，图片文件名为 photo.png。为此，需要在服务器端创建一个支持 Multipart 格式请求数据的 Servlet，我们把 Servlet 命名为 ImageShowerMultipart，Servlet 的代码如下：

```java
package com.ttt.servlet;

import java.io.ByteArrayOutputStream;
import java.io.FileInputStream;
import java.io.FileOutputStream;
import java.io.IOException;
import java.io.InputStream;

import javax.servlet.ServletException;
import javax.servlet.ServletOutputStream;
import javax.servlet.annotation.MultipartConfig;
import javax.servlet.annotation.WebServlet;
import javax.servlet.http.HttpServlet;
import javax.servlet.http.HttpServletRequest;
import javax.servlet.http.HttpServletResponse;
import javax.servlet.http.Part;

import org.apache.tomcat.util.http.fileupload.servlet.ServletFileUpload;

@WebServlet("/ImageShowerMultipart")
@MultipartConfig
public class ImageShowerMultipart extends HttpServlet {
    private static final long serialVersionUID = 1L;

    protected void doGet(HttpServletRequest request, HttpServletResponse response)
throws ServletException, IOException {
        request.setCharacterEncoding("utf-8");
```

```
        String type = request.getParameter("type");
        if ((type == null) || (type.equalsIgnoreCase(""))) {
            type = "1";
        }
        String id = request.getParameter("id");
        if ((id == null) || (id.equalsIgnoreCase(""))) {
            id = "1";
        }

        FileInputStream fis = new FileInputStream(this.getServletContext().getRealPath("") +
                                    "images/png" + type + id + ".png");
        byte[] b=new byte[fis.available()];
        fis.read(b);
        fis.close();

        response.setContentType("image/png");
        ServletOutputStream op = response.getOutputStream();
        op.write(b);
        op.close();
    }

    protected void doPost(HttpServletRequest request, HttpServletResponse response)
                            throws ServletException, IOException {
        request.setCharacterEncoding("utf-8");

        if (ServletFileUpload.isMultipartContent(request)) {
            Part part = request.getPart("image");
            InputStream is = part.getInputStream();

            ByteArrayOutputStream baos = new ByteArrayOutputStream();

            byte[] b = new byte[1024];
            while(is.read(b)>0) {
                baos.write(b);
            }

            b = baos.toByteArray();

            FileOutputStream fos = new
                    FileOutputStream(this.getServletContext().getRealPath("") +
                                    "images/photo.png");
            fos.write(b);
            fos.close();
        }

        doGet(request, response);
    }
}
```

在 Servlet 中，为了处理 Multipart 格式的请求数据，我们为这个 Servlet 加上了 @MultipartConfig 标注。并在 doPost()函数中判断是否为 Multipart 格式的请求数据，若是，则从请求数据包中获得数据流并将数据流保存在这个 Web 工程的 "images/photo.png" 文件中。

现在编写客户端代码。新建一个名为Ex13Network05的Android工程，修改res/layout/activity_main.xml文件，修改后的文件内容如下：

```xml
<LinearLayout xmlns:android="http://schemas.android.com/apk/res/android"
    android:layout_width="match_parent"
    android:layout_height="match_parent"
    android:orientation="vertical">

    <Button
        android:id="@+id/id_btn_1"
        android:layout_width="match_parent"
        android:layout_height="wrap_content"
        android:text="@string/text_btn_1" />

    <Button
        android:id="@+id/id_btn_2"
        android:layout_width="match_parent"
        android:layout_height="wrap_content"
        android:text="@string/text_btn_2" />

    <Button
        android:id="@+id/id_btn_3"
        android:layout_width="match_parent"
        android:layout_height="wrap_content"
        android:text="@string/text_btn_3" />

    <Button
        android:id="@+id/id_btn_4"
        android:layout_width="match_parent"
        android:layout_height="wrap_content"
        android:text="@string/text_btn_4" />

    <ImageView
        android:id="@+id/id_iv"
        android:layout_width="match_parent"
        android:layout_height="match_parent"
        android:scaleType="fitCenter"
        android:contentDescription="@string/hello_world"
    />

</LinearLayout>
```

再修改res/values/strings.xml文件，在其中定义几个在布局文件中用到的字符串引用，内容如下：

```xml
<?xml version="1.0" encoding="utf-8"?>
<resources>

    <string name="app_name">Ex13Network05</string>
    <string name="hello_world">Hello world!</string>

    <string name="text_btn_1">显示第一张蝴蝶</string>
    <string name="text_btn_2">显示第二张蝴蝶</string>
    <string name="text_btn_3">显示第一张卡通</string>
    <string name="text_btn_4">显示第二张卡通</string>
```

```
</resources>
```

现在修改 MainActivity.java 文件，修改后的文件内容如下：

```java
package com.ttt.ex13network05;

import android.graphics.Bitmap;
import android.graphics.BitmapFactory;
import android.net.ConnectivityManager;
import android.net.NetworkInfo;
import android.os.Bundle;
import android.os.Environment;
import android.os.Handler;
import android.support.v7.app.AppCompatActivity;
import android.view.View;
import android.widget.Button;
import android.widget.ImageView;
import android.widget.Toast;

import java.io.File;
import java.io.IOException;

import okhttp3.Call;
import okhttp3.Callback;
import okhttp3.MediaType;
import okhttp3.MultipartBody;
import okhttp3.OkHttpClient;
import okhttp3.Request;
import okhttp3.RequestBody;
import okhttp3.Response;

public class MainActivity extends AppCompatActivity implements View.OnClickListener {
    private ImageView iv;
    private Bitmap bm;
    private Handler handler;

    @Override
    protected void onCreate(Bundle savedInstanceState) {
        super.onCreate(savedInstanceState);
        setContentView(R.layout.activity_main);

        bm = null;
        handler = new Handler();

        iv = (ImageView) this.findViewById(R.id.id_iv);

        Button btn_1 = (Button) this.findViewById(R.id.id_btn_1);
        btn_1.setOnClickListener(this);
        Button btn_2 = (Button) this.findViewById(R.id.id_btn_2);
        btn_2.setOnClickListener(this);
        Button btn_3 = (Button) this.findViewById(R.id.id_btn_3);
        btn_3.setOnClickListener(this);
        Button btn_4 = (Button) this.findViewById(R.id.id_btn_4);
        btn_4.setOnClickListener(this);
    }
```

```java
private boolean checkNetworkState() {
    ConnectivityManager cm = (ConnectivityManager) this
            .getSystemService(MainActivity.CONNECTIVITY_SERVICE);
    NetworkInfo ni = cm.getActiveNetworkInfo();
    if ((ni == null) || (!ni.isConnected())) {
        return false;
    }

    return true;
}

@Override
public void onClick(View v) {
    if (!checkNetworkState()) {
        Toast.makeText(this, "网络没有打开,请打开网络后再试。",
                Toast.LENGTH_LONG).show();
        return;
    }

    int id = v.getId();

    switch(id) {
        case R.id.id_btn_1:
            downloadImageAndShow(1, 1);
            break;
        case R.id.id_btn_2:
            downloadImageAndShow(1, 2);
            break;
        case R.id.id_btn_3:
            downloadImageAndShow(2, 1);
            break;
        case R.id.id_btn_4:
            downloadImageAndShow(2, 2);
            break;
    }
}

private void downloadImageAndShow(final int type, final int id) {
    new Thread(new Runnable() {
        @Override
        public void run() {
            OkHttpClient client = new OkHttpClient();

            MediaType MEDIA_TYPE_PNG = MediaType.parse("image/png");
            RequestBody requestBody = new MultipartBody.Builder()
                    .setType(MultipartBody.FORM)
                    .addFormDataPart("type", ""+type)
                    .addFormDataPart("id", ""+id)
                    .addFormDataPart("image", "photo.png",
                            RequestBody.create(MEDIA_TYPE_PNG,
                                new File(Environment.getExternalStorageDirectory() +
                                    "/photo.png")))
                    .build();

            Request request = new Request.Builder()
```

```
                .url("http://192.168.233.131:8080/" + "ForAndroid/ImageShower
Multipart")
                .post(requestBody)
                .build();

        client.newCall(request).enqueue(new Callback() {
            @Override
            public void onFailure(Call call, IOException e) {
                e.printStackTrace();
            }

            @Override
            public void onResponse(Call call, Response response)
                                                    throws IOException {
                if (!response.isSuccessful()) {
                    return;
                }

                byte[] b = null;
                try {
                    b = response.body().bytes();
                } catch (IOException e) {
                    e.printStackTrace();
                    return;
                }
                bm = BitmapFactory.decodeByteArray(b, 0, b.length);
                handler.post(new Runnable() {
                    @Override
                    public void run() {
                        iv.setImageBitmap(bm);
                    }
                });
            }
        });
    }).start();
}
```

与 Ex13Network04 例子不同的是，我们采用 Multipart 来封装请求数据，代码如下：

```
        RequestBody requestBody = new MultipartBody.Builder()
                .setType(MultipartBody.FORM)
                .addFormDataPart("type", ""+type)
                .addFormDataPart("id", ""+id)
                .addFormDataPart("image", "photo.png",
                        RequestBody.create(MEDIA_TYPE_PNG,
                            new File(Environment.getExternalStorageDirectory() +
                                "/photo.png")))
                .build();
```

在 Multipart 中，我们放置了 3 个请求参数，包括名字为 type 的字符串 String、名字为 id 的字符串 String 和一个文件。

我们还需要修改 AndroidManifest.xml 文件，在其中添加如下的权限申请：

```
<uses-permission android:name="android.permission.ACCESS_NETWORK_STATE"/>
<uses-permission android:name="android.permission.INTERNET"/>
```

运行该程序，即可得到与图 13-3 完全一样的效果。同时，观察服务器端，在 Web 的工作目录下将产生一个用于存储从客户端发来的图片文件的目录。

13.9 同步练习三

使用 OkHttp 编写一个进行网络信息注册的程序，包括客户端程序和服务器端程序，注册的信息包括姓名、出生日期、密码、电话号码和头像。可以使用任何熟悉的方式，由于请求数据中包含头像，因此，只能使用 POST 请求发送数据。

13.10 使用 JSON 格式的数据与服务器端通信

13.10.1 JSON 基础

当前，JSON 已经作为一种数据交换的标准格式广泛应用于数据表示中。那么，什么是 JSON，以及如何使用 JSON 来表示数据呢？本节将对 JSON 的概念和应用进行描述。

JSON 是一种轻量级的数据交换格式，它使用"name/value"对来表示数据。JSON 支持两种结构：以"{}"（大括号）表示的对象数据和以"[]"（中括号）表示的数组数据。JSON 表示数据的能力在于：通过这两种基础格式的组合可以表示任何复杂的数据。例如，为了表示一个人的基本信息，可以使用如下的 JSON 格式的数据表示：

```
{
    "name": "Geoge Bush",
    "age": 20,
    "memo": "乔治毕业于哈佛大学，获得计算机科学博士学位……",
    "phone": "13800138000"
}
```

这段数据表示的信息非常清楚：一个人，他的名字叫作"Geoge Bush"，他的年龄为"20岁"，他的简介为"乔治毕业于哈佛大学，获得计算机科学博士学位……"，他的电话号码为"13800138000"。想要表示多个人的信息，可以使用如下代码：

```
[
    {
        "name": "Geoge Bush",
        "age": 20,
        "memo": "乔治毕业于哈佛大学，获得计算机科学博士学位……",
        "phone": "13800138000"
    },
    {
        "name": "Bill Gates",
        "age": 23,
        "memo": "比尔……",
        "phone": "13800138001"
    }
    ……
]
```

也即使用 JSON 的"[]"来表示多个人的信息。

访问 JSON 数据，根据 JSON 数据的格式不同而有所不同：如果 JSON 数据是对象数据，则使用"变量名.成员名"的方式访问；如果 JSON 数据为数组数据，则使用"变量名[下标].成员名"的方式进行访问。例如，假设我们的第一个 JSON 对象的变量名为 var1，则可以使用 var1.name 的方式访问 name 属性的值；假设我们的第二个 JSON 数组的变量名为 var2，则可以使用 var2[2].name 的方式访问第二个人的 name 属性。

JSON 的属性数据类型可以是任何计算机支持的数据类型，包括数字、字符串、逻辑值（true 或 false）、数组、对象、null。例如，下面是一个更为复杂的 JSON 数据：

```
{
    "name": "Bill Gates",
    "workday": ["Monday", "Tuesday", "Friday"],
    "salary": 8700.5,
    "birth": "1980-10-10",
    "memo": "Something……"
    "alive": true
}
```

13.10.2 在 JavaScript 中使用 JSON 数据

JSON 是 JavaScript 支持的原生数据格式，因此，在 JavaScript 中使用 JSON 数据非常简单。例如，在 JavaScript 中可以直接定义变量的值是一个 JSON 数据，内容如下：

```
var bill =
{
    "name": "Bill Gates",
    "workday": ["Monday", "Tuesday", "Friday"],
    "salary": 8700.5,
    "birth": "1980-10-10",
    "memo": "Something……"
    "alive": true
}
```

进而可以使用"变量名.成员名"或"变量名[下标].成员名"的方式访问 JSON 数据。

13.10.3 在 Java 中使用 JSON 数据

Java 并不是直接支持 JSON 数据的。在 Java 中，任何一个 JSON 数据都被看成一个字符串，称为"JSON 串"。通过使用第三方提供的 Jar 包，可以将 JSON 串转换为一个 Java 的对象，也可以将一个 Java 对象转换为一个 JSON 串。在这些第三方包中，目前较好用且使用较为广泛的 JSON 包是 Google 提供的 gson。可以从 Google 的开发者网站上下载 gson 包。我们使用 gson 的 2.3.1 版本，也即，gson-2.3.1.jar。下载完毕后，将 gson-2.3.1.jar 复制到工程目录的 libs 目录下，然后选择该包并单击鼠标右键，在弹出的快捷菜单中单击"Add as Library"即可使用 gson 包了。

现在我们简单地演示一下如何使用 gson 实现 JSON 串与 Java 对象之间的转换。首先，自定义一个 Java 类，例如，定义为 Person 类，代码如下：

```
public class Person {
    public String name;
    public int age;
    public String memo;
    public String phone;
}
```

基于这个 Java 类，我们可以将一个 JSON 串转换为 Person 对象，以及将一个 Person 对象转换为 JSON 串，代码如下：

```
        Gson gson = new GsonBuilder().setDateFormat("yyyy-MM-dd").create();

        String bill_json =
        "{" +
           "\"name\":" + "\"Bill Gates\"" + "," +
           "\"age\":" + "20" + "," +
           "\"memo\":" + "\"比尔毕业于哈佛大学……\"" + "," +
           "\"phone\":" + "\"13800138000\"" +
        "}";
        Person bill = gson.fromJson(bill_json, Person.class);
        System.out.println(bill.name + "\n" + bill.age + "\n" + bill.phone + "\n" + bill.memo);

        Person geoge = new Person();
        geoge.name = "Geoge";
        geoge.age = 23;
        geoge.phone = "13800138001";
        geoge.memo = "Something to say……";
        String geoge_json = gson.toJson(geoge);
        System.out.println(geoge_json);
```

在这段代码中，我们首先获得一个 Gson 对象，然后将一个表示 Person 数据的 JSON 串转换为 Person 类的对象，最后，再将一个 Person 类的对象转换为 JSON 串。运行这个代码片段，将显示如图 13-5 所示的结果。

```
Bill Gates
20
13800138000
比尔毕业于哈佛大学……
{"memo":"Something to say……","name":"Geoge","phone":"13800138001","age":23}
```

图 13-5　使用 gson 实现 Java 对象和 JSON 串之间的转换

13.10.4　使用 POST 方法及 JSON 数据格式发送请求

下面我们通过一个具体的例子来演示如何使用 HTTP 的 POST 方法和 JSON 数据格式与服务器端程序进行 HTTP 通信。继续使用 13.2 节中的点击按钮显示相应图片的例子。程序运行效果与图 13-3 完全一样。

为了处理 JSON 格式的数据请求，我们在服务器端首先创建一个名为 ImageShowerJSON 的 Servlet，代码如下：

```
package com.ttt.servlet;

import java.io.FileInputStream;
import java.io.IOException;

import javax.servlet.ServletException;
import javax.servlet.ServletOutputStream;
import javax.servlet.annotation.WebServlet;
import javax.servlet.http.HttpServlet;
import javax.servlet.http.HttpServletRequest;
import javax.servlet.http.HttpServletResponse;
```

```java
import com.google.gson.Gson;
import com.google.gson.GsonBuilder;

@WebServlet("/ImageShowerJSON")
public class ImageShowerJSON extends HttpServlet {
    private static final long serialVersionUID = 1L;

    protected void doGet(HttpServletRequest request, HttpServletResponse response)
                                throws ServletException, IOException {
        request.setCharacterEncoding("utf-8");
        Gson gson = new GsonBuilder().setDateFormat("yyyy-MM-dd").create();
        String data = request.getParameter("data");
        TypeAndId ti = gson.fromJson(data, TypeAndId.class);

        FileInputStream fis = new FileInputStream(
            this.getServletContext().getRealPath("") + "images/png" + ti.type + ti.id + ".png");
        byte[] b=new byte[fis.available()];
        fis.read(b);
        fis.close();

        response.setContentType("image/png");
        ServletOutputStream op = response.getOutputStream();
        op.write(b);
        op.close();
    }

    protected void doPost(HttpServletRequest request, HttpServletResponse response)
                                throws ServletException, IOException {
        doGet(request, response);
    }

    private class TypeAndId {
        public int type;
        public int id;
    }
}
```

然后新建一个名为 Ex13Network07 的 Android 工程，并将 gson-2.3.1.jar 复制到工程的 libs 目录下。接下来修改 res/layout/activity_main.xml 文件，文件内容如下：

```xml
<LinearLayout xmlns:android="http://schemas.android.com/apk/res/android"
    android:layout_width="match_parent"
    android:layout_height="match_parent"
    android:orientation="vertical">

    <Button
        android:id="@+id/id_btn_1"
        android:layout_width="match_parent"
        android:layout_height="wrap_content"
        android:text="@string/text_btn_1" />

    <Button
        android:id="@+id/id_btn_2"
        android:layout_width="match_parent"
        android:layout_height="wrap_content"
```

```xml
            android:text="@string/text_btn_2" />

        <Button
            android:id="@+id/id_btn_3"
            android:layout_width="match_parent"
            android:layout_height="wrap_content"
            android:text="@string/text_btn_3" />

        <Button
            android:id="@+id/id_btn_4"
            android:layout_width="match_parent"
            android:layout_height="wrap_content"
            android:text="@string/text_btn_4" />

        <ImageView
            android:id="@+id/id_iv"
            android:layout_width="match_parent"
            android:layout_height="match_parent"
            android:scaleType="fitCenter"
            android:contentDescription="@string/hello_world"
            />

</LinearLayout>
```

修改 res/values/strings.xml 文件，内容如下：

```xml
<?xml version="1.0" encoding="utf-8"?>
<resources>

    <string name="app_name">Ex13Network07</string>
    <string name="hello_world">Hello world!</string>

    <string name="text_btn_1">显示第一张蝴蝶</string>
    <string name="text_btn_2">显示第二张蝴蝶</string>
    <string name="text_btn_3">显示第一张卡通</string>
    <string name="text_btn_4">显示第二张卡通</string>

</resources>
```

现在修改 MainActivity.java 文件，文件内容如下：

```java
package com.ttt.ex13network07;

import android.graphics.Bitmap;
import android.graphics.BitmapFactory;
import android.net.ConnectivityManager;
import android.net.NetworkInfo;
import android.os.Bundle;
import android.os.Handler;
import android.support.v7.app.AppCompatActivity;
import android.view.View;
import android.widget.Button;
import android.widget.ImageView;
import android.widget.Toast;

import com.google.gson.Gson;
import com.google.gson.GsonBuilder;
```

```java
import java.io.File;
import java.io.IOException;

import okhttp3.Call;
import okhttp3.Callback;
import okhttp3.FormBody;
import okhttp3.MediaType;
import okhttp3.MultipartBody;
import okhttp3.OkHttpClient;
import okhttp3.Request;
import okhttp3.RequestBody;
import okhttp3.Response;

public class MainActivity extends AppCompatActivity implements View.OnClickListener {
    private ImageView iv;
    private Bitmap bm;
    private Handler handler;

    @Override
    protected void onCreate(Bundle savedInstanceState) {
        super.onCreate(savedInstanceState);
        setContentView(R.layout.activity_main);

        bm = null;
        handler = new Handler();

        iv = (ImageView) this.findViewById(R.id.id_iv);

        Button btn_1 = (Button) this.findViewById(R.id.id_btn_1);
        btn_1.setOnClickListener(this);
        Button btn_2 = (Button) this.findViewById(R.id.id_btn_2);
        btn_2.setOnClickListener(this);
        Button btn_3 = (Button) this.findViewById(R.id.id_btn_3);
        btn_3.setOnClickListener(this);
        Button btn_4 = (Button) this.findViewById(R.id.id_btn_4);
        btn_4.setOnClickListener(this);
    }

    private boolean checkNetworkState() {
        ConnectivityManager cm = (ConnectivityManager) this
                .getSystemService(MainActivity.CONNECTIVITY_SERVICE);
        NetworkInfo ni = cm.getActiveNetworkInfo();
        if ((ni == null) || (!ni.isConnected())) {
            return false;
        }

        return true;
    }

    @Override
    public void onClick(View v) {
        if (!checkNetworkState()) {
            Toast.makeText(this, "网络没有打开，请打开网络后再试。",
                    Toast.LENGTH_LONG).show();
            return;
```

```java
        }

        int id = v.getId();

    switch(id) {
        case R.id.id_btn_1:
            downloadImageAndShow(1, 1);
            break;
        case R.id.id_btn_2:
            downloadImageAndShow(1, 2);
            break;
        case R.id.id_btn_3:
            downloadImageAndShow(2, 1);
            break;
        case R.id.id_btn_4:
            downloadImageAndShow(2, 2);
            break;
    }
}

private void downloadImageAndShow(final int type, final int id) {
    new Thread(new Runnable() {
        @Override
        public void run() {
            OkHttpClient client = new OkHttpClient();

            Gson gson = new GsonBuilder().setDateFormat("yyyy-MM-dd").create();
            TypeAndId ti = new TypeAndId();
            ti.type = type;
            ti.id = id;
            String json = gson.toJson(ti);

            RequestBody formBody = new FormBody.Builder()
                    .add("data", json)
                    .build();
            Request request = new Request.Builder()
                .url("http://192.168.233.131:8080/ForAndroid/ImageShowerJSON")
                .post(formBody)
                .build();

            client.newCall(request).enqueue(new Callback() {
                @Override
                public void onFailure(Call call, IOException e) {
                    e.printStackTrace();
                }

                @Override
                public void onResponse(Call call, Response response)
                                                    throws IOException {
                    if (!response.isSuccessful()) {
                        return;
                    }

                    byte[] b = null;
                    try {
```

```
                    b = response.body().bytes();
                } catch (IOException e) {
                    e.printStackTrace();
                    return;
                }
                bm = BitmapFactory.decodeByteArray(b, 0, b.length);
                handler.post(new Runnable() {
                    @Override
                    public void run() {
                        iv.setImageBitmap(bm);
                    }
                });
            }
        });
    }
}).start();
}

@SuppressWarnings("unused")
private class TypeAndId {
    public int type;
    public int id;
}
```

我们定义了一个 TypeAndId 类,用于组装发送到服务器端获取指定图片的参数。然后,在 downloadImageAndShow()函数中,我们使用如下代码:

```
Gson gson = new GsonBuilder().setDateFormat("yyyy-MM-dd").create();
TypeAndId ti = new TypeAndId();
ti.type = type;
ti.id = id;
String json = gson.toJson(ti);
RequestBody formBody = new FormBody.Builder()
        .add("data", json)
        .build();
```

获得一个 Gson 对象,并将图片的 type 和 id 封装在 TypeAndId 对象中,然后使用 gson 将数据转换为 JSON 串,并将 JSON 串作为 data 的参数发送给服务器端程序。在运行代码之前,我们还需要在 AndroidManifest.xml 文件中申请相应的权限,代码如下:

```
<uses-permission android:name="android.permission.ACCESS_NETWORK_STATE"/>
<uses-permission android:name="android.permission.INTERNET"/>
```

运行代码,即可得到如图 13-3 所示的结果。